PROOF AND KNOWLEDGE IN MATHEMATICS

PROOF AND KNOWLEDGE IN MATHEMATICS

Edited by
Michael Detlefsen

London and New York

First published 1992
by Routledge
2 Park Square, Milton Park, Abingdon, Oxfordshire OX14 4RN
Simultaneously published in the USA and Canada
by Routledge
a division of Routledge, Taylor & Francis
711 Third Avenue, New York, NY 10017
First issued in paperback 2014

Routledge is an imprint of the Taylor and Francis Group, an informa business

Transferred to Digital Printing 2006

© 1992 Michael Detlefsen

Typeset in 10/12pt Times by
Leaper & Gard Ltd, Bristol

All rights reserved. No part of this book may be reprinted or reproduced or utilized in any form or by any electronic, mechanical, or other means, now known or hereafter invented, including photocopying and recording, or in any information storage or retrieval system, without permission in writing from the publishers.

British Library Cataloguing in Publication Data
Proof and knowledge in mathematics.
1. Mathematics
I. Detlefsen, Michael
511.3

Library of Congress Cataloging in Publication Data
Proof and knowledge in mathematics / edited by Michael Detlefsen.
p. cm.
Includes bibliographical references.
1. Mathematics—Philosophy. 2. Logic, Symbolic and mathematical.
I. Detlefsen, Michael.
QA8.4.P75 1992
511.3—dc20   91-16885

ISBN 978-0-415-06805-5 (hbk)
ISBN 978-1-138-00935-6 (pbk)

Publisher's Note
The publisher has gone to great lengths to ensure the quality of this reprint but points out that some imperfections in the original may be apparent

For my teacher, Arthur F. Holmes

CONTENTS

Notes on contributors		viii
Preface		1
1	PROOF AS A SOURCE OF TRUTH *Michael D. Resnik*	6
2	REFLECTIONS ON THE CONCEPT OF *A PRIORI* TRUTH AND ITS CORRUPTION BY KANT *William W. Tait*	33
3	LOGICISM *Steven J. Wagner*	65
4	EMPIRICAL INQUIRY AND PROOF *Shelley Stillwell*	110
5	ON THE CONCEPT OF PROOF IN ELEMENTARY GEOMETRY *Pirmin Stekeler-Weithofer*	135
6	MATHEMATICAL RIGOR IN PHYSICS *Mark Steiner*	158
7	FOUNDATIONALISM AND FOUNDATIONS OF MATHEMATICS *Stewart Shapiro*	171
8	BROUWERIAN INTUITIONISM *Michael Detlefsen*	208
	Index	251

NOTES ON CONTRIBUTORS

Michael Detlefsen is Professor of Philosophy at the University of Notre Dame and Editor-in-Chief of the *Notre Dame Journal of Formal Logic*. His areas of special interest are the philosophy of logic and mathematics. He is the author of *Hilbert's Program: An Essay on Mathematical Instrumentalism* (Reidel, 1986). Currently, he is working on a book on intuitionism, and co-authoring another with Michael Byrd on Gödel's theorems.

Michael D. Resnik is University Distinguished Professor of Philosophy at the University of North Carolina, Chapel Hill. Author of *Frege and the Philosophy of Mathematics* and *Choices*, as well as many articles in logic and the philosophy of mathematics, he is a leading exponent of the view known as mathematical structuralism.

Stewart Shapiro received his B.A. from Case Western Reserve University in 1973 and his Ph.D. from the State University of New York at Buffalo in 1978. He is currently an Associate Professor of Philosophy at the Ohio State University at Newark. He has published several articles in logic and philosophy of mathematics, and edited a collection of essays entitled *Intensional Mathematics* (North-Holland, 1985). His most recent book is entitled *Foundations without Foundationalism: A Case for Second-Order Logic* (Oxford University Press, 1991).

Mark Steiner was born in 1942, received the A.B. degree in 1965 from Columbia University, and the Ph.D. degree in 1972 from Princeton University. He taught at Columbia from 1970 to 1977, and subsequently at the Hebrew University of Jerusalem where he is Associate Professor of Philosophy. He is author of *Mathematical*

NOTES ON CONTRIBUTORS

Knowledge (Cornell University Press, 1975) and a number of articles on the philosophy of mathematics.

Pirmin Stekeler-Weithofer, born in 1952, has an M.A. in Linguistics (1975) and Mathematics (1977), and a Ph.D. and *venia legendi* in Philosophy (1984 and 1986 respectively). He is a lecturer in the Philosophy Department of the University of Konstanz in Konstanz, Germany. His main areas of research are the philosophy of logic and mathematics and the philosophy of language. He is the author of *Grundprobleme der Logik: Elemente einer Kritik der formalen Vernunft* (Walter de Gruyter, 1986).

Shelley Stillwell was educated in the United States at Kenyon College (A.B.) and Purdue University (Ph.D.). Currently, she is on the faculty of Illinois State University. Her research interests include the later Wittgenstein's epistemology and philosophy of mathematics, confirmation theory, contemporary epistemology, and modal logic.

William W. Tait received his Ph.D. from Yale University and is currently Professor of Philosophy at the University of Chicago. During the first part of his career, he worked in proof theory with the aim of establishing constructive foundations for significant parts of classical mathematics. Towards the end of the 1960s, technical considerations convinced him that not even a seriously restricted fragment of classical analysis could be constructively founded. He also became convinced by Wittgenstein's *Philosophical Investigations* that constructive reasoning has no special epistemological status. Because of this, he abandoned proof theory and began to think about the foundations of mathematics in terms of the gap between the notion of truth and the notion of proof-on-the-basis-of-generally-accepted-principles. His current project concerns Zermelo's conception of set theory, which sees it as an enterprise that always goes beyond any particular axioms and points to new truths.

Steven J. Wagner received his A.B. in philosophy from the University of California, Berkeley, in 1973, and his Ph.D. in philosophy from Princeton in 1978 (advisor, Paul Benecerraf). He has taught philosophy at the University of Illinois, Urbana-Champaign, since 1977. He was also a visiting professor at the University of Washington, Seattle. His work centers on classical problems of philosophy – truth, objectivity, mind and body, platonism, and the nature of justification. A book, *Truth, Pragmatism and Ultimate Theory*, is in progress. He is the co-editor (with Richard Warner) of *Beyond Naturalism and Physicalism* (Blackwell, forthcoming).

PREFACE

The essays in this volume all deal with questions which arise in the course of trying to arrive at an epistemology for mathematics. Prominent among these are questions concerning the nature of justification in mathematics and its possible sources. Some of the essays in this volume treat these questions in a general way, while others focus on various more particular concerns relating to them. Among these more particular concerns, perhaps the most noteworthy are the following: (i) the question of the *a priori* versus *a posteriori* character of mathematical justification (including the question of the role of proof as a source of warrant in mathematics), (ii) the question of the character of mathematical reasoning or inference, especially the question of what role (if any) logic has to play in it, and (iii) the question of the epistemological importance of the formalizability of proof.

In "Proof as a source of truth," Michael Resnik addresses the general question of mathematical justification and its sources by taking up the question of how, if at all, a proof comes to provide a warrant for its conclusion. His argument makes use of the general framework developed in his earlier work on pattern recognition, in which mathematical objects are treated as patterns, and patterns as abstractions of material (and hence perceivable) things called "templates." In his view, proofs are to be seen as concerned with patterns, both establishing (as conclusions) and utilizing (as premises) laws that are structurally isomorphic to laws that have been verified for templates. In this way, templates become the gateway through which we gain epistemic access to the objects of mathematics dealt with in (realist) proofs.

This same question, albeit in its more familiar guise as that concerning the possible *a priori* character of mathematical

1

justification (and the oft-associated question concerning the necessity or contingency of the propositions thus justified), is also a focus of concern of the papers by William Tait, Steven Wagner, Shelley Stillwell, Pirmin Stekeler-Weithofer, and Mark Steiner. Tait, in his "Reflections on the concept of *a priori* truth and its corruption by Kant," gives a historical discussion of these issues. His particular concern is to construct a reading of Kant which separates his views on the *a priori* character of both pure *and* applied geometry from a certain narrow conception of logical inference that has been (wrongly, in Tait's view) attributed to him. Thus, though he does not defend Kant's conception of geometry, he nevertheless finds its chief weakness to reside in something other than its conception of logic.

In "Logicism," Steven Wagner admits that we use empirical data in justifying belief in mathematical truths, but holds that this is so only because we suffer such deficiencies of cognitive capacity as lack of memory and limited power of concentration. For ideal agents not suffering such limitations, knowledge of mathematical truths could (would?) be *a priori*. Wagner also argues that mathematics (or, at least, arithmetic and set theory) becomes necessary in the sense that any rational being characterized by such things as a desire to understand, a dissatisfaction with recognizedly limited and incomplete theories, and a need to generalize and unify his beliefs will have reason to develop (some version of?) it.

Shelley Stillwell's "Empirical inquiry and proof" takes an even stronger position. She not only denies that we must or do use empirical data in epistemizing mathematical proofs; she also claims that we *cannot* do so. Her argument is a development of the Wittgensteinian idea that what proof does is to provide standards for what it is that is to count as a permissible description of our experience, and that that is something that experience itself cannot do.

Stekeler-Weithofer takes a view opposed to the above defenders of the *a priori* character of the warrant provided by proof. In "The concept of proof in elementary geometry," he argues that geometrical proofs rely on observational knowledge: we know, he says, nothing of space or geometrical form save what we know empirically about the possibilities of moving bodies from one place to another. Nongeometrical proofs, too, depend upon empirical knowledge since, were we not able to perceive figures and characters of language in an intersubjectively valid and stable way, we

PREFACE

would not be able to use syntactical and semantical rules in a uniform way, and were we to lack the ability to do that, proof would be impossible. Therefore, proof – both of the geometrical and the nongeometrical varieties – is dependent for its epistemic force on empirical knowledge.

Mark Steiner's essay "Mathematical rigor in physics" takes a still different approach. He begins by noting some striking facts concerning the application of mathematics to physics; namely, that accuracy in physics has been improved not only through the application of mathematical theories outside the domain of their validity, but also through the use of inconsistent theories! He (rightly, in my opinion) takes this to be a datum of potentially great importance to the philosophy of mathematics, and raises the (recognizably Kantian) question of whether it can be accounted for without positing some kind of "pre-established" harmony between the laws of the behavior of physical bodies and the laws of thought.

The general question of the nature of mathematical reasoning and, in particular, the place of logic in it is a common theme of the papers by Stewart Shapiro, Wagner, and myself. Shapiro and Wagner both maintain (though for rather different reasons) the importance of logical inference in mathematical proof, while I argue to the contrary.

Wagner's idea is basically that logic – and, specifically, first-order, classical logic – captures the inferential patterns present in an ideal scheme of mathematical justification. His view is that mathematics is essentially a structure of proofs – proofs which consist in deductions from axioms and represent ideal lines of justification. (The axioms themselves, however, are not justified by proof, but rather by explanatory inference from intuitions). The importance of (first-order, classical) logical inference to mathematical reasoning is that it captures the inferential structures figuring in this scheme of ideal justifications.

In "Foundationalism and foundations of mathematics," Stewart Shapiro also argues for the importance of logic to mathematical proof, though he is primarily concerned not with specific answers to the question of which logic is most plausibly described as the logic of mathematics, but rather with the larger question of how candidates for such an office are to be selected. One of these larger questions has to do with basic conceptions of what the task of logic is supposed to be. Here Shapiro distinguishes two basic conceptions of logic, the rationalist and the semantical. The former

3

sees a system of logic as designed to codify the ideally rational inferences of a given language; the latter sees it as intended to clarify the division between inferences that hold regardless of the subject-matter with which one is concerned and those which owe their validity to the peculiarities of a given subject-matter. He also distinguishes two different kinds of completeness questions; namely, (i) whether a given formalism codifies every correct inference expressible in the language of the formalism, and (ii) whether a given formalism codifies every mathematically correct inference. Classical first-order logic is an example of a formalism that is complete in the first sense, says Shapiro. He believes, however, that its widespread popularity as an instrument for formalization is largely the result of certain unfortunate foundationalist impulses – chief among which is an overestimation of the certainty that is required for mathematical knowledge – but that first-order languages cannot be the basis of formalisms complete in the second sense. To capture correct mathematical reasoning, he argues, we need a stronger second-order logic. And since it is sheer foundationalist propaganda that mathematical knowledge requires the security or certainty that restriction to classical first-order logic brings with it, we should have no epistemological qualms about the move to second-order reasoning.

In my "Brouwerian intuitionism," I offer a considerably more deflationary view of logic, arguing that (because of its supposedly topic-neutral character) logical inference is not well suited for use in genuinely mathematical reasoning. This raises the challenge of formulating and defending an alternative conception of rigor which does not depend upon a rigorous inference's being emptied of all substance, but it also permits one to make the necessary distinction between the epistemic condition of the genuine mathematician and that of the mere logical master.

These same considerations cause me to reject the idea that mathematical proof must be formalizable. Stillwell, too, rejects this idea, arguing (*pace* Wittgenstein) that proofs are not construction-tokens, but rather construction-types; that is, that they do not represent a kind of experimentation requiring apprehension of sensible features of spatio-temporal concretia (which, on one view, is what formal proofs are), but rather an introduction of new concepts which requires for its grasp an appreciation of its conceptually creative significance.

Wagner takes the contrary view, arguing for the formalizability

PREFACE

of proof in the sense that formalized proof constitutes an ideal for mathematical justification. Though actual mathematical practice may often (indeed typically) fall short of this ideal, the epistemic worth of actual mathematical reasoning is nonetheless a function of the extent to which it approximates this ideal. Hence, formalizability is an important part of what gives proof its epistemic force on Wagner's view.

The above remarks are in no way intended to serve as a summary of the essays appearing here; and they give nothing like an adequate impression of the richness of the essays, either collectively or individually. Nor are they in any way intended to suggest – what is manifestly false – that the essays cannot profitably be read in a more selective way. They are rather intended to indicate just a few of the topical connections which exist among the essays in this volume, and thus to suggest a few of the many ways in which they might be read as a body.

With the exception of my "Brouwerian intuitionism," the essays here are appearing for the first time. My thanks also to Brian Rosmaita, who corrected the proofs and compiled the index for this collection. I would also like to thank the authors for their hard work and generous spirit of cooperation. It has been a pleasure to work with them.

Michael Detlefsen
Notre Dame, Indiana
December 1990

1

PROOF AS A SOURCE OF TRUTH

Michael D. Resnik

SUMMARY

For mathematical realists an especially pressing question is that of how a proof, for example of Euclid's theorem, can establish a conclusion about mathematical objects, for example that there are infinitely many prime numbers. In trying to give a realist answer to this question, I interpret it as the question of how a proof of, say, Euclid's theorem could (i) induce us to believe that there are infinitely many prime numbers and (ii) give us good reasons for so believing. My answer to the last question is that our proof of Euclid's theorem induces us to believe that there are infinitely many primes, because we have been prepared through our mathematical education to understand its component statements and to follow its reasoning. We have also been prepared, through learning an appropriate background theory, to see the proof as presenting, arranging and transforming information about numbers, and hence have also been prepared to draw from it the conclusion that there are infinitely many prime numbers. That takes care of the belief-inducing part of my answer. The good reasons part is trickier. We want to know why the proof of Euclid's theorem gives us good reasons for believing that there are infinitely many primes. Now viewed from our own standards of reasoning it is clear that Euclid's proof gives us good reasons for believing its conclusions. We need only check the proof to see that it conforms to our standards of rigor. But for our present purposes this is not enough. Everything is based upon accepting an "appropriate background theory." We need assurance that such a theory is not just an elaborate mythology passed from generation to generation by the priesthood of mathematics. Towards this end I turn to speculative history about the development of ancient mathematics, hoping thereby to give a rough idea of the sort of "background theory" that connects proofs with mathematical objects and through comparing its genesis with developments in science to dispel the mythological clouds that some might see clinging to it.

PROOF AS A SOURCE OF TRUTH

I. INTRODUCTION

Taken literally and seriously, mathematics affirms truths about numbers, functions, sets, spaces and other entities, which are as real as rocks and yet inhabit neither space-time nor our minds. There are good reasons, which I shall not consider here, for philosophers of mathematics to take mathematics seriously and literally.[1] Opposing those reasons is the ensuing mystery such realism makes of mathematical knowledge. If numbers and their cousins are outside of space-time, then they cannot transmit information to our sensory detectors. If they are also neither individual nor collective mental constructions, then we are not free to imagine or stipulate or otherwise "dream up" their properties. How then can we acquire knowledge about them? That question drives mathematical realists to despair and makes reluctant nominalists of many sensible people.

A special case of this question is the truth/proof problem. It may be formulated as follows: when mathematicians prove mathematical statements they construct diagrams, write formulas, and produce arguments. On the face of it, none of these spatio-temporal or mental transactions could provide them with information about the abstract world of mathematical objects. Thus how can proving a mathematical statement show that it is true? For that matter, how can anything mathematicians do produce information about mathematical objects?[2]

To assess the magnitude of this problem compare proving in mathematics with observing in, say, biology. To find out whether a certain species of fish carries its eggs internally until its young hatch and swim free, biologists might capture members of the species, provide them with a suitable breeding ground and observe whether they bear their young live. Although our biologists might encounter serious problems arranging for this experiment and the results might be inconclusive, there is no serious question that observing the fish provides information about them. One reason why there is no serious question about this is that physiologists and psychologists have developed (rudimentary) theories of how information about events in fish tanks is transmitted to our brains. On the other hand, our methods for acquiring mathematical knowledge seem to contrast sharply with perception, and we do not have even rudimentary scientific theories concerning the mechanisms whereby we learn about the mathematical realm.

This essay attempts to solve the truth/proof problem. Section II formulates a more tractable version of the problem. Sections III and IV are concerned with proofs, how they work and what they achieve. Section V begins with a brief and incomplete summary of my view of mathematical objects as positions in patterns. This opens the way for a historically grounded (but still hypothetical) account of how the study of numerical patterns may have linked the subject-matter of mathematics with proofs.

II. REFORMULATING THE TRUTH/PROOF PROBLEM

Formalists and intuitionists do not have a truth/proof problem. For them a mathematical statement is true just in case it is provable, and proofs are syntactic or mental constructions of our own making. Their example suggests that we might solve the problem by closing the gap between ourselves and mathematical reality, but it also shows the danger of sacrificing mathematical realism in the process. Thus my first task will be to formulate a version of mathematical realism which still produces a genuine truth/proof problem.

The version I have in mind is immanent realism about mathematical objects.[3] With other realist views, it holds that mathematical objects are abstract entities existing independently of us and our constructions and theories. It also holds that (most of) the claims of contemporary mathematics are true, and that they are true independently of our holding them to be true. Thus immanent realism about mathematical objects is incompatible with formalism, intuitionism, deductivism, and the other familiar antirealist views in the philosophy of mathematics. Despite its title, immanent realism affirms that mathematical reality transcends our own existence, beliefs, and experience. Immanent realism derives its title from its *immanent conception of truth*. It uses a conception of truth that applies only to sentences within its own language, whereas transcendent versions of realism employ conceptions of truth that transcend their home language through applying to sentences in a variety of languages.

More precisely, the field of application of the immanent realists' truth-predicate is restricted to their (our) own language. They take truth and reference as merely disquotational: Using a classical metalanguage, they accept instances of the usual Tarski disquotational schemata,

PROOF AS A SOURCE OF TRUTH

"p" is true if and only if p,
"N" denotes x if and only if $x = N$,
x satisfies "P" if and only if Px,

and the laws relating the denotation, satisfaction, and truth-conditions of compound expressions to those of their components. Immanent realists rest content with a Tarskian definition of truth based upon list-like specifications for the references of primitive terms.[4] These specify references for primitive terms of the object language by using the same terms within the metalanguage. For example, in specifying the references for the primitive terms of first-order number theory immanent realists might use the following definition:

t refers 1 to y if and only if t = "0" & $y = 0$;
t refers 2 to $\langle x, y \rangle$ if and only if t = "S" and Sxy;
t refers 3 to $\langle x, y, z \rangle$ if and only if t = "+" and $x + y = z$
 or t = "X" and xX$y = z$.

More importantly, unlike their transcendent kin, immanent realists seek no additional word-world theory, such as the causal theory of reference, stated in terms applicable to all languages and by means of which one might determine the reference of a foreigner's term or even that of an unknown term of one's own language.

Within the immanent realists' framework one can formulate the familiar doctrines that have traditionally divided mathematical realists from antirealists. For example, in arguing about whether there are unprovable mathematical truths, realists and antirealists can find material enough in our own language without needing to turn to sentences in arbitrary languages. On the other hand, forgoing a universal theory of truth forces immanent realists to abandon some beliefs often identified with realism. These include, for instance, the belief that mathematicians, no matter what language they speak, aim to construct true theories, and the belief that if there is a Martian mathematics then it affirms many of the truths that we do. Despite this, immanent realism with respect to mathematical objects is still saddled with all the traditional objections to mathematical objects, for it still maintains that they exist outside of space–time and are independent of us and our theorizing. In particular, immanent realism still faces the truth/proof problem.

For both immanent and transcendent realists this problem comes to the worry of how proving something, say Euclid's prime

9

number theorem, establishes that it is true. But to the immanent realist, to establish that Euclid's prime number theorem is true is just to establish that there are infinitely many primes. More generally, to establish that p is true is just to establish p. So for the immanent realist, the truth/proof problem amounts to the problem of explaining why p is provable only if p.

But this generates a new problem – the problem of giving the truth/proof problem a nontrivial and *philosophical* interpretation. One could read the truth/proof problem as the mathematical task of proving the soundness of our rules of proof. But this cannot be the correct way to understand the problem, because even formalists could recognize such a soundness proof as a bona fide mathematical result and construe it in their own terms. On the other hand, turning from mathematical technicalities could lead us to philosophical trivialities. It is tempting to say, "Well, it's just part of the meaning of *proof* that one cannot prove p unless p."

The way out of this dilemma is to remember where we began. Our original problem was to show how activities involving reasoning, paper and pencil calculation, drawing diagrams, and the like could establish anything about the mathematical realm. Immanent realists still face this problem. But they do not have to deal with an additional problem of explaining how some nontrivial, truth-making relation manages to obtain between provable mathematical sentences and mathematical reality.

Proving has an ontic dimension for most constructivists; to them proving p establishes its truth in the sense of *making* p true. But for us it does not; proving p establishes, shows, or demonstrates p only in the *epistemic sense* of providing us with good reasons for believing p. This suggests that we interpret the truth/proof problems as this pair of questions:

1   In view of the gap between us and mathematical reality why does proving p induce us to believe p?
2   Why are the reasons a proof provides good reasons?

For many of us a glib answer to both questions may apply. This is the response that in our mathematics courses we are trained to accept proofs as giving good reasons, and we have been conditioned to believe things which we think we have good reasons to believe. Some, influenced perhaps by Wittgenstein, would interject that the glib answer applies to all of us – practicing mathematicians and uninspired mathematics students alike. Mathematics, on their

PROOF AS A SOURCE OF TRUTH

view, is not a science, and there are no mathematical facts. There is nothing but a certain social practice in which proving plays a major role. Thus, they would continue, the question we should ask is: how did our *practice* of proving mathematical statements evolve? That is a good question too, and answering it will take us a large way towards answering the two questions I posed above. However, from my realist perspective, no answer to those two questions can be fully satisfactory unless ultimately we find some connection between proving and mathematical objects. But first let us take a closer look at proving.

III. HOW PROOFS ARE USED AND WHERE THE TRUTH/PROOF PROBLEM ARISES

Consider some of the ways in which mathematicians use proofs. They are used, of course, for demonstrating new results, but also for giving alternative demonstrations of previous results. Moreover, under that heading we can distinguish, on the one hand, proofs which show that a previous result can be given a weaker or more economical demonstration (as in replacing a nonconstructive proof by a constructive one), and, on the other hand, proofs that use methods from a different area of mathematics to obtain a known result (as in the algebraic proofs of the completeness theorem). Proofs are also used for presenting axiomatic derivations of previously obtained unsystematized results (as in the Dedekind–Peano axiomatization of number theory) as well as for reformulating extant axiomatic theories (as Hilbert, Veblen, and others did for geometry). In addition, one might prove that one mathematical statement is equivalent to or implies another (as in investigating the consequences and equivalents of the axiom of choice), and one might test or justify a proposed definition by proving that it has certain consequences (as in Frege's derivation of the laws of arithmetic from his definition of number).

Some will maintain that when we develop the consequences of axioms or definitions or other mathematical hypotheses we are simply applying logic and that we are likely to appeal to something beyond logic only when we prove nonlogical statements in an unaxiomatized branch of mathematics. I am inclined to think that this view is wrong and that even axiomatic investigations involve more than just using logic as a theorem generator. But we need not take up that issue now, since we can focus on proofs that attempt to demonstrate new results in unaxiomatized areas.

A popular view is that no result has been fully demonstrated until it has been derived from an accepted set of axioms within an accepted formal system. On that view most historical proofs in unaxiomatized fields such as number theory, algebra, and analysis should be regarded as incomplete proof sketches – much as philosophers of science used to consider the explanations one actually finds in scientific writings to be explanation sketches. For brevity let us call such informal, unaxiomatized demonstrations *working proofs*. Contemporary mathematics is replete with working proofs. Even as methodologically conscious a field as mathematical logic developed unaxiomatized, and in teaching it virtually every logician presents it as a collection working proofs and theorems. Plainly these proofs are no less convincing psychologically than fully formalized proofs, and few think that we need to axiomatize mathematical logic in order to secure its results.

Of course, epistemologists might claim that this is irrelevant. Conceding that practicing mathematicians are right to rest content with or even to insist upon working proofs, they might go on to say that a proper understanding of our knowledge of mathematical results depends upon specifying the axioms implicit in their proofs. Frege may have made a more radical claim: some of his remarks imply that we do not really know theorems until they have been properly and formally proved. However, the epistemologists I have in mind are less radical; they claim that axioms are somehow implicit in working proofs despite the unwillingness or even inability of mathematicians to supply them.

I will grant that someone very skilled in logic and axiomatics could be expected to produce a reasonable axiomatization of any collection of related mathematical results.[5] But what would this show? Not that this was the true basis for mathematicians' knowing the results in question, but at best that it *could* be the basis for somebody's knowing them. Such a possibility might have some bite, if the axioms in question could be shown to be knowable *a priori* or if the resulting axiomatic proofs could be shown to be more certain or reliable than working proofs. However, bitter experience has shown that both of these conditions are false. Axiomatic proofs do not guarantee the truth of their theorems.

To synthesize and systematize a significant body of results we need general axioms. Sometimes we find that a very plausible axiom is too general and must be restricted to exclude counterexamples. An infamous case is the naive set-theoretic axiom to the effect that every condition defines a set. Sometimes we find that we need an axiom that is neither intuitively plausible nor perspicuous. Russell and Whitehead conceded this of their axioms of infinity and reducibility. We used to say the same of the axioms of choice and replacement, although with time they grew on us. In sum, recent history teaches us that, rather than being acclaimed as self-evident truths already hidden in our proofs, new axioms are likely to be greeted with skepticism and accepted only gradually. Axioms may come and go while the central theorems of a branch of mathematics, even those with difficult and unintuitive proofs, remain on the books.

These facts do much to undercut the belief that axiomatic proofs make their results more certain or reliable. It also shows that in practice axioms need not carry the familiar quasi-psychological marks of the *a priori*, such as being self-evident, indubitable, or thought with necessity and universality. Nor is it plausible that the typical axiom is *a priori* in the sense of being a necessary condition for the possibility of any thought at all.[6] Of course, it is easily argued with respect to many axioms that they are necessary conditions for certain types of mathematical thinking to be possible. For instance, if we dropped the associativity axiom for addition over the real numbers, then our ability to reason about the reals would be severely constricted. But this does not show that the associativity axiom is *a priori*, since the same point is true of any central theoretical principle, even those in biology and economics.

The history of the failed attempts to reduce mathematics to logic and Quine's critiques of conventionalism and the analytic–synthetic distinction give us additional grounds for concluding that proofs from so-called *a priori* axioms do not provide the true path to mathematical knowledge. Let us see where that leaves us. We want to know how proving, say, Euclid's theorem establishes it – not in the sense of making it true, but rather in the sense of causing us to believe it and also in the sense of giving us good grounds for believing it. Furthermore, we want to explain how a proof of this theorem could be convincing to one who holds that it is true just in case there are infinitely many prime numbers. We have also decided to focus on working proofs of new results, which need not be even implicitly axiomatic. The next question for us to consider is how working proofs work.

PROOF AND KNOWLEDGE IN MATHEMATICS

IV. HOW WORKING PROOFS WORK

Let us examine a simple proof of Euclid's theorem. Please pretend that this is a working proof of a new result. Try to forget that the theorem has been known for centuries and that the proof that I shall give derives from one presented in the oldest and best known axiomatic treatise.

We want to prove that infinitely many natural numbers are primes. By definition, a prime number is one that can be divided without remainder by only 1 and itself. Thus 2, 3, 5, 7, and 11 count as primes, but 6 which equals 2×3 does not.

We have already established that there are some primes [1], so it suffices for us to show that there is no greatest prime. To this end let suppose that there is one and derive a contradiction. Let p_k be the greatest prime. Let the finite list of primes in increasing order be

$$p_1, p_2, \ldots, p_k \ [2]$$

and consider the number $E = (p_1 \times p_2 \times \ldots \times p_k) + 1$ [3]. This number is greater than p_k [4], and so it cannot be prime. Hence we can divide E into factors until we eventually reach prime factors [5]. Let these be $q_1, q_2, \ldots, q_m$. Then E is a product of these qs. Each q divides E without remainder and no p does; hence no p is identical to any q. It follows that the qs are primes not in our list of primes. This is a contradiction. QED.

How does this proof succeed in convincing us of its conclusion? First, it presents us with certain data. I have indicated each datum above with a bracketed numeral; more explicitly they are as follows: [1] there are some primes; [2], [3] if a greatest prime exists, the primes may be listed and multiplied; [4] the successor of the product of these primes is greater than each of its factors; and [5] any nonprime may be factored entirely into primes. Second, the proof transforms and reformulates some of the data in a way that prompts us to draw new conclusions from it. For instance, by representing E in the form

$$(p_1 \times p_2 \times \ldots \times p_k) + 1,$$

we are led to see that none of the primes p_i can divide it, and thus none of them is among its prime factors. Finally, the proof arranges the data and the various formulations of it so as to prompt us to see that we cannot accept the data and deny the theorem without involving ourselves in a contradiction.

Organizing and formulating ideas and information are extremely important in mathematics. This is well illustrated by our theorem and its proof, as we can see by reformulating both. First, let us reformulate the idea of the infinitude of the primes. Instead of saying that there is no greatest prime, let us say that for each finite collection of primes there is a greater prime. Next let us replace our indirectly organized proof with a directly organized one. Given some primes $p_1, p_2, \ldots, p_k$, we now use the guts of the old proof to establish the existence of an even greater prime which is itself less than or equal to the number E. (I leave the details as an exercise.)

Our proof has psychological, social, and logical features. It has a psychological task of convincing its readers of its conclusion. Its notation and the way and order in which its ideas are presented play important roles in determining whether the proof succeeds at this task. Obscurely written proofs, no matter how rigorous, do not do this job as well as perspicuous ones. From the logical point of view, our proof also has the task of placing its audience in a position to defend its conclusion by means of a cogent argument.

Since people sometimes defend a belief based upon a proof by rehearsing that very proof, one might suspect that a proof's ability to provide us with a defense for a belief is not a logical feature of the proof but just a psychological one. One might suspect that defending our beliefs by giving proofs might just be a very refined form of brain washing, that we do not argue logically but just cleverly dupe our audience by rehearsing a proof. We can quickly quash that suspicion by recalling that a logically impeccable proof, which can thus be used to defend its conclusion, may be very obscure and convincing only to those who reflect on it long and hard (personal example – most proofs of the Schroeder–Bernstein theorem). Conversely, some mistaken proofs have been very effective in convincing their audiences. Thus I think there is a clear separation between the psychological and logical powers of a proof.

Finally, proofs are social and cultural objects. They evolve in a particular social and cultural context, and they have intended audiences. Our proof of Euclid's theorem will not convince someone who has no grasp of English or who has never been exposed to mathematical reasoning. It is important to underscore the social and cultural features of proofs, because we might forget that part of finding out how a proof works includes finding out how its intended audience (the author included) came to be prepared to follow it.

Proofs are context bound. This point plainly applies to a proof's language and notation, but it also applies to its reasoning and data.[7] At the beginning of the nineteenth century proofs in analysis freely used geometrical diagrams and reasoning; the end of that century saw geometry banned from analysis in favor of arithmetic and set theory. Bolzano and Dedekind tried to prove that infinite sets exist by arguing that any object of thought can be thought about and thus give rise to a new thought object. Today we reject such proofs and use an axiom of infinity. Newton, Leibniz, and their disciples calculated derivatives and integrals by reasoning about infinitesimals, a method replaced about a hundred years ago by epsilon–delta style proofs. Ironically, today some say that nonstandard analysis is proof that Newton and Leibniz were right all along. What counts as data in a given branch of mathematics varies from context to context too. To our ancestors, who did not recognize the imaginary or negative numbers, it was absurd to think that every number has a square root or that one could subtract one number from another without worrying about producing as much nonsense as results from dividing by zero. On the other hand, Cantor and Frege blithely took it for granted that they could specify sets almost at will, and Dedekind managed to characterize the natural number sequence while unconsciously running together subclass and membership.

Surveying these ramblings, we see that working proofs work by presenting information in a way that is psychologically convincing and rationally defensible. In order for them to work, however, they must be tailored to their audience, which must be prepared to understand them and follow their reasoning and be willing to take as given the data upon which they are based. There is no way to guarantee that a proof will work for all audiences.

The question we should now ask is: how does the intended audience for a proof become prepared for it? Also we must not forget the question which has been driving us all along: how does a proof show us anything about mathematical objects? As you may have guessed, my answers to these questions are related. Proofs make claims about mathematical objects. *To be able to understand a proof we must be able to understand claims about mathematical objects. If we are prepared to do that then we will be prepared to*

learn truths about mathematical objects from proofs.

Understanding claims about mathematical objects requires training, and understanding advanced mathematical claims requires much more training than that required to grasp the first principles of arithmetic. In this respect learning mathematics is like learning science. We use simple experiments to confront school children directly with elementary facts about the behavior of water at various temperatures or the effects of magnets on each other. On the other hand, it takes us years of training, including training in mathematics, to understand theoretical science. Moreover, our contact with highly theoretical entities is very indirect and mediated by many layers of theory.

We should not expect that all or even most mathematical activity, anymore than other scientific activity, will present mathematical objects to us directly, in any reasonable sense of directness. To solve the truth/proof problem we need not show that every proof at every level of mathematics directly presents us with mathematical objects. I do think, however, that a solution should show us how giving proofs in elementary mathematics does present us with mathematical objects. If one can do that, then the balance of the solution will be to explain how contact with mathematical objects at the elementary levels suffices to produce information-yielding proofs in advanced mathematics as well. In the next section I shall sketch a view of the nature of mathematical objects and attempt to clarify the sense in which elementary mathematics makes direct, information-producing, contact with them.

V. CONTACT WITH MATHEMATICAL OBJECTS

The nature of mathematical objects determines how we make contact with them. My view is that mathematical objects are positions in patterns with the geometrical point being my paradigm mathematical object. It is a featureless atom, with no properties or identifying features beyond those it has by virtue of its relationships to other geometrical objects. Now I say that the same holds for other mathematical objects. We talk of sets and sequences, for instance, as if they were composite; but, on my view, sets are not literally made up of their members – they are simply related to them by membership. I have elaborated this view elsewhere, and we need not consider its details here.[8] The feature of it that is crucial here is that we can never study a mathematical object, such as pi, in isolation, as we might study a specimen. Instead we study the patterns or structures in which mathematical objects are positions.

You might think that I would now argue that we make perceptual contact with patterns, at least simple finite ones, because we can see, hear, and feel them. This would give us a basis for the rudimentary mathematical knowledge we need to build the larger edifice. But mathematical structures are as abstract as their positions, on my view. Both numbers and numbers systems are causally inert, so there is no more reason to think that we see patterns with our senses than there is to think that we see numbers. (We do speak, to be sure, of seeing the pattern that several things fit, but that is a matter of seeing that they are arranged in such and such a way.)

We can learn about patterns, however, by using our senses to observe an arrangement of things that fit them. Of course, we cannot compare the pattern and the arrangement in question in order to see whether the latter fits the former. But we can see whether the arrangement satisfies conditions that specify the pattern in question or at a more rudimentary level we can compare the arrangement with a concrete representation of the pattern. For instance, to see whether a sample road sign conforms to the pattern for STOP signs, we can examine the sample to see whether it satisfies a standard description of STOP signs or we can lay a standard STOP sign next to it and compare them. I will speak of *templates* when referring to such concrete representations of patterns, and I will speak of concrete things *conforming* to a template in order to distinguish that from their fitting an abstract pattern. Templates need not instantiate the patterns they represent; written musical scores, for example, are templates for sound patterns.

I have little idea of how humans came to use templates or how the highly sophisticated conventions governing their use evolved. The process is probably similar to or part of the evolution of language. Nor do I know how we, as members of an advanced culture, pick up the trick of templates. However, there is no question that we have the ability to construct and use them and exercise it almost daily. Furthermore, I see no reason to think that we must account for this ability or the practice of using templates in terms of our having some prior knowledge or grasp of abstract entities of any kind. Rather I think that the direction is the other way around: first we evolved the practice of using templates; then we posited abstract patterns corresponding to them. The direction is important here, because we cannot perceptually compare a putative instance of a pattern with that pattern. Nor can we carry out the comparison at one step removed; we cannot compare patterns with their templates in order to check the accuracy of the latter. On the other hand, we can often perceptually compare some templates with concrete things purporting to conform to them.

Another thing I do not claim to understand is how we can learn about a template by examining things that conform to it – for example, how we can learn about the engineering drawings for the 1976 Chevy Nova by examining the heap my daughter drives. This too is something we developed along with our practice of using templates, although, obviously, it is no simple matter to learn what to ignore when determining whether something conforms to a template or how to infer features of a template by inspecting just one instance. Again it is enough for my purposes that we recognize that we have this ability and that it does not seem to involve knowing abstract patterns either.

I think that templates are the key to understanding the development of mathematical knowledge. There is historical and anthropological evidence that well before a culture develops anything we would regard as mathematics its members draw geometric designs and diagrams and develop devices for counting and tallying the things counted. Although I think that geometric templates played as important a role in the development of mathematics as numerical templates, in the remainder of this essay I will focus on the latter.

At the most primitive level people use bags of pebbles or tally sticks to count and record counts.[9] In seeking efficiency they will quickly come to use special pebbles or marks for key tally points, just as we mark through four slashes to signal each fifth item on a tally sheet. Further refinements of this system might lead us to aggregate our marks in rows containing, say, at most ten fifth tally symbols. Then each filled row would indicate a total of fifty items counted. The Roman numeral system seems to be an outgrowth of an approach like this. On the other hand, we need not use special symbols for counts of 5, 10, etc. if we simply arrange our tally marks in patterns. For instance instead of writing

PROOF AND KNOWLEDGE IN MATHEMATICS

when counting 1 through 6, we might write

• •   • •   • •   • • •
• •   • •   • • •   • • •

which is not only more compact but also suggests how the numbers are related to one another, as we shall see shortly. Another good feature of this change from the first linear dot notation is that we have preserved the cardinality of each group of dots and thus can re-represent any information recorded by the former in a more perspicuous form.

Because our current decimal notation uses different simple symbols for 0 through 9, it is easier to read and write than either the dot or the Roman notations. It also works immeasurably better in calculating, because it is a place notation, that is, one in which both the digit and the place it occupies within a compound numeral carry information. But while our notation is better suited for store keepers, it is less suited for fledgling number theorists. Let me illustrate this by doing a bit of early Greek mathematics.

The Greeks classified numbers by shape terms. Thus some numbers were triangular, others were square, others still were oblong. Suppose I represent 1 through 6 as follows:

1   2   3   4   5   6
•   • •   •   • •   • •   • • •
      • •   • •   • • •   • • •

Then 1 counts as both a degenerate square and triangle, 2 as oblong, 3 as triangular and 4 as square. At this point one might ask which numbers are triangular, which are squares, which are oblong. To this end let us write down sequences of their dot representations and see whether some patterns emerge. Starting with the oblong numbers we have

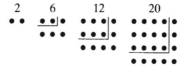

Each oblong pattern is generated from the previous one by adding on a column of dots on the right that ends with a new row of dots added to the bottom. If you look at these examples you should see

PROOF AS A SOURCE OF TRUTH

that we start with an array one dot high and two dots wide, move to one two dots high and three dots wide, and then on to one three dots high and four dots wide, so that in general the nth array will be n dots high and $n+1$ dots wide and can be obtained from the previous one by adding n dots arranged vertically next to the right edge and then n dots along the bottom edge. Thus we have the following result:

(a) The nth oblong array (or number) contains $n(n+1)$ dots.

But also notice that since the nth array is obtained by adding $2n$ dots to the previous array and we start with a two-dot array, the nth array contains as many dots as the sum of the first n even numbers. Thus we also obtain:

(b) The sum of the first n even numbers equals $n(n+1)$.

If we express this in modern notation as

(c) $2 + 4 + 6 + \ldots + 2n = n(n+1)$,

we can easily prove it by mathematical induction. The Greeks did not know this principle, but I find their dot proof just as convincing.[10]

It is interesting how easily our attention has shifted. We started out by asking for an arithmetical characterization of the oblong numbers and ended with a formula for the sum of the first n even numbers. This suggests that we now ask two sorts of questions: What is the formula for numbers of a given shape? What sums do they represent? This will quickly lead us to use the dot technique to find that the square numbers are characterized by the formula n^2 and that

(d) $1 + 3 + 5 + \ldots + 2n - 1 = n^2$.

What about the first n numbers? Well, each oblong array contains as many dots as the sum of the first n even numbers; so if we divide these arrays in half by drawing diagonals through them and erasing all the dots below the diagonal, then each new triangular array will have half as many dots and can be obtained from the previous one by adding half the dots to it as were necessary to obtain the oblong from which it was produced. This means that the nth triangular array will contain $\frac{1}{2}(n(n+1))$ dots and will come from the previous array by adding n dots to it. This yields the formula

(e) $\quad 1 + 2 + 3 + \ldots + n = \frac{1}{2}(n(n+1))$.

Again we can easily verify this independently by mathematical induction. But we can also derive it more quickly from (c), since (c) may be rewritten as

(f) $\quad 2 \cdot 1 + 2 \cdot 2 + 2 \cdot 3 + \ldots + 2n = n(n+1)$,

from which (e) follows by dividing both of its sides by 2. It should take you but a moment to realize that this method for obtaining (e) can be generalized to yield a formula for the sum of the first n multiples of any number p, namely

(g) $\quad 1p + 2p + 3p + \ldots + np = \frac{1}{2}(pn(n+1))$.

Our modern arithmetical notation gave us an advantage over the Greeks and allowed us to quickly find (g). But the Greeks could have arrived at the same result by reasoning about dots. Suppose that each dot in the dot triangles used in the derivation of (e) is replaced by p dots. Since we have increased the dots in each array p-fold, the nth array of dots will represent the sum of the first n multiples of p. Since it contains $p[\frac{1}{2}(n(n+1))]$ dots, this number also equals the sum of the first n multiples of p.

In terms of mathematical development our excursion has jumped centuries ahead of the time when people were struggling to improve methods for recording tallies. We have implicitly assumed that dot patterns can be extended indefinitely and that cardinality is preserved under rearrangements of dots. I have also let us quietly slip from talk about concrete tokens of tally symbols and templates into talk about numbers and abstract patterns. I have done so purposely in order to let you see how naturally focusing on templates for counting can lead one to full-fledged mathematics. Let us now return to the early historical period to see if we can trace these transitions more closely.

At the same time that people developed systems for tallying, they probably began to record empirically derived laws relating tallies to tallies. I have in mind laws such as "two fives make ten" and "four fives make twenty." In much the manner of John Stuart Mill's account of arithmetic, we can think of these laws as implicitly referring to the results to be expected from aggregating counted collections. For example, "two fives make ten" would be shorthand for "counting five things and then counting five more things will yield a final count of ten things." It is also likely that, just as they do in our culture, these sentences did double duty as laws or rules of calculation, allowing one to rewrite, say, "5 + 5" as "10."

Counting cows and jugs of wine may have given rise to the law that $5 + 5 = 10$, but as a prediction about the results of actually counting cows or jugs of wine, this law is too easily refuted. Between the time we count the two fives and the time we count the ten somebody might steal some jugs of wine or one of the cows might calf. On the other hand, dots in templates are much more stable and much easier to survey and manipulate. Thus it probably came to pass that ancient pre-mathematicians began to use their numerals to refer directly to standard tally marks and only indirectly to things tallied. Furthermore, they probably also hardened the normative force of their rules of calculation in the sense of ruling out templates that failed to conform to them just as we exclude from ordinary English expressions not composed from our alphabet, such as " >}#@@." These steps would have represented great progress towards giving mathematics an abstract subject-matter and insulating it from the vicissitudes of the world of everyday experience, but take note that neither abstract numbers nor abstract patterns would be in place yet.

Once our mathematical ancestors had developed principles of calculation, they would have been in a position to pose and solve simple "mathematical" problems containing unknowns, such as, "what cipher yields the same result whether we add it to itself or multiply it by itself?" Much of our knowledge of Eygptian and Babylonian mathematics comes from writings containing worked out sets of problems of this sort. Evidently, the Babylonians also gave the first proofs, since they would check or prove the solution to an equation by substituting it in the equation and determining whether the result was correct. Their proofs used their principles of calculation as data.[11]

I have been interpreting the rudimentary mathematics of our hypothetical ancestors in formalist and nominalist terms. Under this interpretation these people had no real truth/proof problem. In asking what x is such that $x + 1 = 2x - 4$, they are asking a question about numerical templates. To fix our ideas, let us suppose that their templates were constructed of dots arranged in linear forms. Let us suppose that to add two templates they simply juxtaposed them. To substract a smaller template from a larger one they would line up the smaller one below one end of the larger and then cross off every pair of vertically aligned dots. Finally, to multiply two dot templates they would lay one horizontally and the other vertically so that their end-points overlapped. Then they would fill in the space below the horizontal dots until they obtained a rectangular array, which they would then rearrange in the standard linear form to yield the product of the original templates.

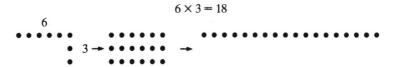

If this is how our ancestors added, subtracted, and multiplied, then in asking what x is such that $x + 1 = 2x - 4$, they were asking what templates are such that if we add one dot to it, then we get the same result as we would if we doubled it and then subtracted four dots. Of course, if they expressed themselves in a notation like ours, the connection between the assertion "$5 + 1 = 2 . 5 - 4$" and dot templates would not be utterly transparent but it could be easily recovered from the conventions governing their notation.

Using the dot system I have just described one can quickly come to appreciate that addition and multiplication are associative and commutative, that multiplication distributes over addition and that 1 is a multiplicative identity. Indeed, this dot system provides a much better understanding of why those principles hold than anything we are usually given.

We are now back at the point where we left the Greeks. Let us consider how they might deal with the truth/proof problem. Some might have used the nominalist solution that we just presented, but almost certainly others regarded numbers as abstract entities. Let us call them platonists. It is likely that the platonists thought of numbers as abstract dot patterns, that is types, corresponding to the dot templates we have been considering until now. If so, then the platonist solution to the truth/proof problem would run along the following lines. Proofs are concerned with the patterns that our mathematical templates represent. These templates and the operations on them are structurally isomorphic to the patterns they represent, and hence laws governing templates may be projected onto their associated patterns. Furthermore, proofs of theorems

PROOF AS A SOURCE OF TRUTH

concerning patterns are based upon images of laws which have been verified for templates. Thus these projected laws and the theorems we derive from them must be true of patterns. Finally, we can often check such conclusions independently by verifying that templates obey the laws corresponding to a theorem proved about patterns.

This platonist account not only tells us how some proofs can yield information about mathematical patterns but also how we make contact with mathematical objects. We do it through templates for mathematical patterns. Now some people are sure to protest that this hardly deserves to be called contact with mathematical objects. Besides, it looks as if I have still left us with the big mystery of how one can know that templates and their associated patterns are isomorphic when we only have perceptual access to the former.

I will give a quick solution to the mystery. Patterns are *posits*. Our hypothetical ancestors introduced them and gave them the status of abstract entities in order to meet their need for infinite analogs to templates, such as the natural number sequence, and for geometrical ideals, such as dimensionless points, one-dimensional lines and perfect circles. Although I am not prepared to argue that ancient mathematicians consciously posited abstract patterns, today explicitly positing new entities frequently leads to groundbreaking advances in mathematics and physics, and it is not unheard of in the other sciences either.[12]

There is a bit more to my quick solution, because merely positing new entities is never enough to advance science or mathematics. One must relate the entities already on hand with the new ones in order to tie the latter to the experiences and problems which led to their positing. Also one must explain how we can obtain information about the new entities. These are issues to be addressed when the entities are posited and to be solved only by postulating that they have specific properties which are sufficient for connecting them with us and the old entities.

My remarks about positing and postulating are *epistemic*, not ontic. I am not claiming that in positing mathematical or scientific objects we make or create them, as we might make a tool or create a painting. Nor am I claiming that in postulating that, say, every point on a circle is equidistant from its center, we endow circles with that property much as we might be said to endow a boat with a color by painting it. Circles, numbers, neutrinos, and quarks existed and obeyed their laws long before we posited them. Positing and postulation are simply part of the means by which we discovered this.

Positing fits well with immanent realism. One might ask, "I know that we justifiably posited electrons, but are there really electrons? Is it true that they exist?" To a transcendent realist this might be a challenge to chase after evidence, other than our ordinary sort of evidence for electrons, which shows that electron theory "corresponds to reality." But to the immanent realist the question simply amounts to saying, "I know that we justifiably posited electrons, but are there electrons?" To take this question seriously is to take it epistemically – to take it as asking whether we are still justified in believing in electrons.

Our ancient platonists posited ismorphisms to forge the links between patterns and their previous knowledge of templates. By appealing to the isomorphism between certain templates and certain patterns, mathematicians may project some template properties onto certain abstract patterns. This is not to say that they may project properties of finite templates onto infinite patterns or properties of imperfect templates onto patterns made up of geometrical points, lines, or circles. Nor does this mean that they may directly project properties observed to hold for a few similar templates onto all patterns of a species.[13] No doubt mathematicians do generalize from templates to a whole species of patterns. Even our dot proofs depend upon this type of inference. But the inference involves two steps, one of projection from templates to patterns, followed by one of generalization.

That ends my quick solution, but I still must address the objection that learning about mathematical entities through projection from templates hardly counts as being directly presented with them. One way for me to reply to this is to argue that the mathematical case is not particularly different from the physical case. We know that we use projective reasoning to infer the properties of certain physical objects, distant ones for instance, and it is certainly arguable that we subconsciously do the same for those that are nearby. Now would I be the first to suggest that everyday physical objects are also posits and that using sensations to discern their properties appeals to an implicity postulated connection between the two.[14]

I endorse this reply but I also want to present an additional one. To introduce it I will compare my view with Penelope Maddy's.[15]

PROOF AS A SOURCE OF TRUTH

Like me Maddy emphasizes the analogy between mathematics and the other sciences, and like me she believes that certain rudimentary parts of mathematics form its analogy to the observational base of science. A solution to the truth/proof problem analogous to mine is implicit in her epistemology. She, too, thinks that our contact with mathematical objects at the rudimentary levels of mathematics suffices for building the full edifice. But then we split. Maddy thinks that mathematics and science are parallel disciplines with equally fundamental but separate observational bases. She also maintains that we make direct contact with certain simple mathematical objects – we perceive them with our senses just as we perceive ordinary physical bodies. By contrast I see mathematics and the rest of science as inseparably intertwined. Epistemically, I see both as growing from the trunk of commonsense experience. Because of this I discern no need for a separate observational base for mathematics or for observable mathematical entities. Since for me mathematical entities are theoretical objects, I see no need for our contact with mathematical objects to be more direct than it is with most other scientific entities. With naked eyes we no more observe atoms, cells, small crystals, genes, or even the details of the planets than we observe abstract mathematical patterns. Furthermore, the conclusions we draw from instrument-aided observations implicitly appeal to theories which join the instruments, the objects observed via them, and the readings we obtain. It is true that we invariably develop alternative types of instruments for observing physical objects of a given kind. For instance, we now use both optical and radio telescopes as well as satellite-mounted television cameras to observe planets. But mathematical patterns are no exception, since we invariably find them in a variety of independently given guises and representable by independently derived templates.[16]

VI. BEYOND DOT PROOFS

My account is incomplete. I have not indicated how one might move from the study of simple patterns to advanced mathematics. So I will conclude with a sketch of how that might take place.

Earlier I claimed that numbers, functions, and sets are positions in patterns. So far, however, I have been using dot patterns to represent individual numbers. This suggests, of course, that numbers are *patterns* rather than *positions* in them. It is quite possible that early mathematicians regarded numbers in this way, but as mathematics advanced its attention focused on the entire number sequence rather than on the individual numbers. One infinite pattern replaced the infinitely many separate patterns that were the early numbers. It is easy to see how this transition might have taken place. For if we rearrange triangles, squares, and oblongs of dots as lines we find that each number contains all the earlier numbers as subpatterns.

```
  1     2      3       4        5
  •    • •   • • •   • • • •  • • • • •
```

It is an easy step from here to an unending pattern of dots, the natural number sequence, which contains occurrences of each of the old number patterns. The next move is to think of numbers ordinally, as positions in this sequence, rather than cardinally as collections of dots. Our previous knowledge of numbers is easily recovered, since the nth number is the last member of a collection of n dots.

This new way of representing numbers is sure to lead to new mathematical concepts and theorems. For example, by going back and forth between the old and new representations, one can easily discover facts relating the ordering of numbers to addition and multiplication. Extending the natural number pattern in various ways could lead one to new systems of mathematical objects such as the negative numbers.

However, staring at dot patterns has its limits. It could (and probably did) lead the Greeks to conceive of prime numbers as *indivisible numbers*, since the odd primes are neither square nor oblong. But it is unlikely that staring at dot patterns led them to conceive of primes as the *ultimate factors* of a number or to Euclid's proof that there are infinitely many primes. Not only is the distribution of primes in the natural number sequence irregular, but also the individual primes themselves conform to no geometrical pattern transformable from one prime to yield the next. Two is oblong, three is triangular, and five is trapezoidal (as is every greater odd number).

Our concept of a prime number as an ultimate factor must have emerged from patterns that arise while calculating. Once we begin to list factors of numbers the prime number concept is inevitable. So, too, is the conjecture that every number is prime or may be completely factored into primes. To prove this we only need the

PROOF AS A SOURCE OF TRUTH

definition of prime number and the result that every composite number is a product of finitely many smaller numbers. We can allay any doubts about the latter by remembering that the natural number sequence begins with a first number and that, except for the number 1, a number can be the product of only earlier numbers. (A much harder theorem states that the prime factors of a number are unique up to their ordering in the product. Again a few examples would quickly reveal that pattern, but the proofs of it are not easy, and so discovering one probably had to await further mathematical development.)

This shows us that mathematicians look for patterns even when operating at one or more steps removed from the patterns which form its current subject-matter. They factor some numbers into primes. A pattern emerges: the factorization process leads to smaller and smaller factors until we reach primes. They conjecture that the same holds for all numbers. Returning to the examples, they extract general properties of the numbers and multiplication to use as data for a proof that presents the general factorization pattern exhibited by the examples.

Practical and scientific experiences have often suggested new patterns and generated new branches of mathematics. Mathematics also produces new patterns and theories by reflecting on its own activities. Patterns arise in proofs, calculations, and solutions to problems. Describing them in terms of their positions leads to new theories of new mathematical objects, such as the theories of equations, proof theory, the theory of computability.[17] The process can go on and on.

VII. SUMMARY

The course of this paper has been more than a bit circuitous, so let me conclude by summarizing it. I began with the question:

> How can a proof establish a truth about mathematical objects?

I then introduced the doctrine of immanent realism and used it to transform my initial question to the question:

> How does a proof, for example the above proof of Euclid's theorem, establish a conclusion about mathematical objects, for example that there are infinitely many prime numbers?

PROOF AND KNOWLEDGE IN MATHEMATICS

I interpreted this as the question of how a proof of, say, Euclid's theorem could (i) induce us to believe that there are infinitely many prime numbers and (ii) give us good reasons for so believing.

My answer to the last question ran along these lines. Our proof of Euclid's theorem induces us to be believe that there are infinitely many primes because we have been prepared through our mathematical education to understand its component statements and to follow its reasoning. We have been prepared to evaluate its inferences as well as the data upon which it is based. We have also been prepared, through learning an appropriate background theory, to see the proof as presenting, arranging, and transforming information about numbers, and hence have also been prepared to draw from it the conclusion that there are infinitely many prime numbers.

That takes care of the belief-inducing part of my answer. The good reasons part is trickier. We want to know why the proof of Euclid's theorem gives us good reasons for believing that there are infinitely many primes. Now viewed from our own standards of reasoning it is clear that Euclid's proof gives us good reasons for believing its conclusions. We need only check the proof to see that it conforms to our standards of rigor. But for our present purposes this is not enough. Everything is based upon accepting an "appropriate background theory." We need assurance that this is not just an elaborate mythology passed from generation to generation by the priesthood of mathematics. That is why I turned to speculative history. I hoped thereby to give a rough idea of the sort of "background theory" that connects proofs with mathematical objects, and intended my comparisons with developments in science to dispel the mythological cloud that some might see clinging to such a theory.[18]

NOTES

1 For some discussion of these reasons, see Benacerraf (1973), Field (1980), Putnam (1971), and Quine (1960).
2 The truth/proof problem is implicit in Benacerraf (1973) but the title is taken from Tait's "Truth and proof: the platonism of mathematics" (1986).
3 Unfortunately, I cannot do justice to this view here. I have developed a fuller account in "Immanent realism," currently an unpublished manuscript.

4 Tarski used list-like specifications of primitive reference in his truth-definitions. Field subsequently subjected them to philosophical criticism. See Field (1972).
5 In *Mathematical Knowledge* Steiner (1975) calls such a person a logician-mid-wife.
6 Kitcher (1986) discusses this conception of apriority in connection with mathematics in his "Frege, Dedekind and the philosophy of mathematics."
7 Kitcher (1983) has argued most of these points at length in *The Nature of Mathematical Knowledge*.
8 See my "Mathematics as a science of patterns: ontology and reference" (1981), and "Mathematics from the structural point of view" (1988).
9 See Wilder (1968).
10 See Heath (1981) for more discussion of dot proofs.
11 See van der Waerden (1963).
12 In "Mathematics from the structural point of view" (1988) I present a hypothetical history in which our mathematical ancestors try to describe both the potential infinity of numerical templates and geometrical ideals in terms of possible concrete ones. I conclude that there can be no purely nominalist-modal extension of actual concrete templates to comprehend geometrical ideals, namely perfect circles, straight lines, etc. The reason is that geometrical ideals cannot be identified with specific imperfect concrete approximations, no matter how refined they might be, and to identify them with infinite sequences of possible concrete templates would entail embracing sequences *qua* abstract entities. This impasse led my hypothetical ancients to take the bold step of introducing abstract patterns, and it justified their doing so.
13 I more carefully examine the inferential connection between concrete structures and their abstract images in "Computation and mathematical empiricism" (forthcoming).
14 See Quine (1960).
15 See Maddy (1980).
16 Herein may lie a major reason for our prizing alternative proofs of the same theorem.
17 Although Saunders Mac Lane would not endorse my views or use my terms, his excellent account of the development of mathematical ideas captures the process I have in mind. See his *Mathematics: Form and Function* (1986).
18 I want to thank Michael Detlefsen, Geoffrey Sayre McCord, and Stewart Shapiro for their comments. I read earlier versions of this paper to the Philosophy Department of the University of Rochester and to a conference on the foundations and philosophy of mathematics sponsored by the History and Philosophy of Science Department of the University of Cambridge. I am also grateful to the members of these audiences for their comments and discussion.

REFERENCES

Benacerraf, P. (1973) "Mathematical truth," *Journal of Philosophy* 70: 661-80.
Field, H. (1972) "Tarski's theory of truth," *Journal of Philosophy* 69: 347-75.
—— (1980) *Science without Numbers*, Princeton, NJ: Princeton University Press.
Heath, T. (1981) *A History of Greek Mathematics*, New York: Dover Publications.
Kitcher, P. (1986) "Frege, Dedekind and the philosophy of mathematics," in L. Haaparanta and J. Hintikka (eds) *Frege Synthesized*, Dordrecht: Reidel.
—— (1983) *The Nature of Mathematical Knowledge*, New York: Oxford University Press.
Mac Lane, S. (1986) *Mathematics: Form and Function*, New York: Springer-Verlag.
Maddy, P. (1980) "Perception and mathematical intuition," *Philosophical Review* 89: 163-96.
Putnam, H. (1971) *Philosophy of Logic*, New York: Harper & Row.
Quine, W.V. (1960) *Word and Object*, Cambridge, MA: MIT Press.
Resnik, M. (1981) "Mathematics as a science of patterns: ontology and reference," *Nous* 15: 529-50.
—— (1988) "Mathematics from the structuralist point of view," *Revue International de Philosophie* 42: 400-24.
—— (forthcoming) "Computation and mathematical empiricism," *Philosophical Topics*.
Steiner, M. (1975) *Mathematical Knowledge*, Ithaca, NY: Cornell University Press.
Tait, W. (1986) "Truth and proof: the platonism of mathematics," *Synthese* 69: 341-70.
van der Waerden, B. (1963) *Science Awakening*, New York: Wiley.
Wilder, R. (1968) *The Evolution of Mathematical Concepts*, New York: Wiley.

2

REFLECTIONS ON THE CONCEPT OF *A PRIORI* TRUTH AND ITS CORRUPTION BY KANT*

William W. Tait

> I thought I should take refuge in *logoi*, and study in them the truth of the things that are.
>
> (*Phaedo*, 99d5)

SUMMARY

The distinction is drawn and discussed between a conception of *a priori* truth, which is first found in Plato and is found in Leibniz under the heading of "*a priori* truth," according to which it is truth about a species of structure and can be understood and studied independently of whether or not this kind of structure is exemplified in the natural world, and the conception of "*a priori*" in Kant and later writers, according to which propositions may be *a priori* true of the empirical world.

I

Naively, at least, it seems clear that mathematical truths are *a priori*, since the only warrant that we ultimately accept for a mathematical proposition is a proof and not what is true of the empirical world. But this view is challenged from several directions in the contemporary literature.

One challenge is that, much less being *a priori* true, many theorems of mathematics are not true at all, namely those which imply the existence of "abstract" objects such as numbers and sets, since there are no such objects. This challenge is based on the view that there is a ground, external to mathematics, upon which to judge the truth of the latter. This view seems deeply rooted, and the so-called realism/antirealism issue is still regarded as meaningful by many who write about the philosophy of mathematics and science. For these writers, a mathematical or physical theory might be adequate in all respects that we could possibly test – for example, in the latter case, it might account for a wide range of phenomena in an entirely simple and elegant way – but it might be wrong because it commits us to objects that do not "really exist."

In "Truth and proof: the platonism of mathematics" (1986a) I attempted to undermine the usual picture underlying this view: here is the sensible world of tables and chairs. It gives meaning to the terms in which we speak of it and, once we have learned how to name things, it determines which of the sentences we can express of it are true. Now, here is the language of mathematics and its primitive truths. If this language is to be really meaningful and the primitive truths really true, there must be something like the sensible world which gives meaning to the terms of this language and so determines the truth values of mathematical propositions. My argument was to attack the way meaning and truth are understood in this picture (the "Augustinean" picture, as Wittgenstein called it) in the case of sensible things as well as in the case of mathematics. The realism/antirealism issue is a hydra of confusions and no doubt many heads remain. But I do not want to discuss this here.

Quine challenges the apriority of mathematics and, indeed, the notion of apriority itself in another way.[1] He argues that there is no absolute distinction between the *a priori* and the *a posteriori*; but, rather, there is a spectrum of kinds of propositions, extending from reports of direct observations, through statements of theoretical sciences, to statements of mathematics and logic. Our science is a network of statements which confronts the empirical world as a whole. The position of a sentence in the spectrum is to be understood, not in terms of the source of its truth or falsity or in terms of the source of our knowledge of it, but in terms of our degree of willingness to give up belief in it, as opposed to alternative sentences, when empirical observation demands some adjustment in our science. We are usually more willing to give up a belief based on a single observation than an established physical theory or a principle of mathematics or logic.

Since Kant, at least, the concept of apriority has generally been regarded to be an epistemological concept, concerning the sources of knowledge. This has sometimes been transformed into the claim that it applies primarily to states or processes of knowing rather than to propositions and that the notion of an *a priori* truth is the derivative notion of a proposition that can be known *a priori*. By implication at least, this is the view taken by Bertrand Russell in *A Critical Exposition of the Philosophy of Leibniz*.[2] In *The Nature of Mathematical Knowledge* (1983: chs 1–5), Philip Kitcher challenges the apriority of mathematics by arguing that, if we attempt to analyze the notion of an *a priori* state or process of knowing, we arrive at a concept which is not convincingly applicable to mathematical knowledge.

Russell writes that Kant's equation of apriority with necessity is the result of "confounding sources of knowledge with grounds of truth."[3] He writes, "There is no doubt a great difference between *knowledge* gained by perception, and *knowledge* gained by reasoning; but that does not show a corresponding difference as to what is known" (1937: 24). Presumably, Russell has in mind cases such as "A or not-A," which we can know as a logical truth but, alternatively, might know because we have empirically verified one of the disjuncts. Similarly, in *Naming and Necessity*, Saul Kripke rejects the equation of apriority with necessity on the grounds that one concerns source of knowledge and the other concerns source of truth. He argues that, in fact, neither necessity nor apriority implies the other. I know *a priori* that the standard meter bar is one meter long, but that is not a necessary truth (1980: 54–5). Conversely, "Hesperus = Phosphorus" is necessary, because the names denote rigidly, but not *a priori* (pp. 102–5).

Kripke also offers an example of a proposition which can be known both *a priori* and *a posteriori*. Namely, we can know *a priori* that a certain number is prime because we have proved it. But we can know that it is prime *a posteriori* because a computer proved it (p. 35). Here our knowledge that it is prime is based on our knowledge of the empirical laws of evolution of the states of the computer. Of course, this example presupposes that apriority attaches to states of knowing rather than directly to propositions. But, in any case, the example is puzzling. Assuming that I am a less reliable computer than the machine I use to test the number for primeness, why should the result of my calculations result in *a priori* knowledge and that of the machine in *a posteriori* knowledge? In either case, the conclusion that the number is prime is based on assumptions about the computer – whether it is the machine or me.

What is probably at issue here is a Cartesian notion of "knowledge that," which is a direct insight into truth. Somehow, my computation yields for me this insight and the machine's computation does not. I would question this. But then, I would question the notion of "knowledge that" – at least to the extent that this notion does not admit of degrees, so that I must either know that A or not know that A. Even when A is a theorem of arithmetic, for example, and there is a proof of it which I have "learned," when do I really know that A? When is knowing the proof something more than just having memorized a certain configuration of symbols? I think that "knowing that," propositional knowledge, ultimately reduces to knowing how, competence – and it admits of degrees. In the case of the theorem of arithmetic, I may be able to give a proof on demand but have a greater or lesser command of applying the same principles of proof in other cases, have greater or lesser ability to give examples to show why the particular conditions in the statement of A are essential to its truth, etc. But this, too, I do not want to discuss here.

II

Kant is responsible for confusions about the meaning of apriority, in part because he introduces the notion as applying to cognitions (*Erkenntnisse*) – for example, in his definition at B2–3. But he immediately thereafter gives two criteria (which he assumes are each necessary and sufficient) for *a priori judgements*, namely necessity and strict universality. So judgements must be among the things counted as cognitions. This is corroborated in §19, where Kant is discussing the notion of a judgement. At B141 he defines a judgement to be "nothing but the manner in which given cognitions are brought to the objective unity of apperception." But, in the discussion preceding this definition, he gives two criticisms of the logicians' definition of a judgement as "the representation of a relation between two concepts": first, the definition does not specify what the relation is and, second, it overlooks the case of hypothetical or disjunctive judgement, where the relata are judgements and not concepts. So it is clear that the cognitions to be brought to objective unity in a judgement include concepts (in the case of categorical judgements) and component judgements (in the case of hypothetical and disjunctive judgements). Since it is to judgements, concepts, and intuitions that Kant primarily applies the term "*a priori*," it is likely that these are the items that he was including under the term "*Erkenntnisse*" in his characterization of apriority. So there seems to be little grounds for interpreting him to mean a state or process of knowing by this term. Anyway, since he explicitly equates apriority with necessity, it is surely unreasonable to read him in such a way that the equation becomes obviously false.

III

It is true that Kant's equation of apriority with necessity – and with strict universality – is oddly formulated:

> First, then, if we have a sentence which in being thought is thought as *necessary*, then it is an *a priori* judgement; and if, besides, it is not derived from any sentence except one which also has the validity of a necessary judgement, it is an absolutely *a priori* judgement.... Secondly, ... If, then, a judgement is thought with strict universality, that is, in such a manner that no exception is allowed as possible, it is not derived from experience, but is valid absolutely *a priori*.
>
> (B3–4)

It might seem from this that apriority resides for Kant, not in what is being asserted – the proposition – but in the propositional attitude – the judging – and so is entirely subjective. But Kant does not intend this. In the first place, he did not have our distinction between the proposition and the judgement, at least not explicitly. What corresponds most closely to our notion of proposition is his notion of a problematic judgement[4] and, to our notion of judgement, his notions of assertoric and apodeictic (necessary) judgement. The locus of all judgements is the understanding, and so to say that a judgement is thought as necessary is simply to say that it is necessary. Nor does the necessity rest entirely on our intention. "The apodeictic proposition (*Satz*) thinks the assertoric as determined by these laws of understanding, and therefore as affirming *a priori* ..." (B101). The same true proposition (in our sense of the term) may be judged problematically, assertorically or apodeictically; but these are different judgements. And, to be truly judged apodeictically, their proposition must *in fact* be "determined by" the laws of understanding. As a matter of fact, Kant cannot quite mean this since, for example, the necessary judgement that there is just one line determined by two distinct points is "determined" not by the laws of understanding but by pure intuition. But the important point is that the ground of necessity for a necessary judgement must in fact obtain for the judgement to be true.

IV

The equation of apriority with necessity was not original with Kant; nor did he himself regard his distinction between apriority and aposteriority – in contrast with his distinction between analytic and synthetic judgements – to be original: he wrote in 1790 that "it is a distinction long known and named in logic" (Bk VIII, 228). In particular, Leibniz also equated apriority with necessity.[5] But there is some difference in how they understand the notion of necessity and, indeed, some difference in the scope of what they take to be *a priori* truths. For Leibniz, the *a priori* or necessary truths are those derivable from primitive truths of reason, whose opposite involves an express contradiction. Thus, the necessity of *a priori* truths would seem to be logical necessity for Leibniz. The quote above from 1790 suggests that Kant also takes the concept of apriority, that is, necessity, to be a logical notion, but it is not clear what this can mean for him, since not all *a priori* truths are logical truths. He recognizes two kinds of logic, formal logic and transcendental logic, and what is logically true in either sense is for him *a priori*. But he explicitly excludes the truths of pure geometry from being logical truths in either sense (B14 and B81), although they are *a priori*.[6] On the other hand, applied geometry and the Principle of Sufficient Reason are synthetic *a priori* for Kant but are not *a priori* for Leibniz.[7]

V

Leibniz's definition of *a priori* truth is echoed by Frege. Thus, Leibniz writes in the *Monadology*:

> 33. There are also two kinds of *truths*, those of *reasoning* and those of *fact*. The truths of reasoning are necessary and their opposite is impossible; the truths of fact are contingent, and their opposite is possible. When a truth is necessary, its reason can be found by analysis, resolving it into simpler ideas and simpler truths until we reach the primitives.
> 34. This is how the speculative *theorems* and practical *canons* of mathematicians are reduced by analysis to *definitions*, *axioms* and *postulates*.

THE CONCEPT OF *A PRIORI* TRUTH

35. And there are, finally, *simple ideas*, whose definition cannot be given. There are also axioms and postulates, in brief, primitive principles, which cannot be proved, and which need no proof. And these are *identical propositions*, whose opposite contains an explicit contradiction.

(1948)

In *The Foundations of Arithmetic* Frege writes

The problem then becomes, in fact, that of finding the proof of the proposition, and of following it up right back to the primitive truths. If, in carrying out this process, we come only on general logical laws and definitions, then the truth is an analytic one ... if ... its proof can be derived exclusively from general laws, which themselves neither need nor admit of proof, then its truth is *a priori*.

(1974: 4)

Moreover, in a footnote on page 3 to his discussion of the distinctions between *a priori* and *a posteriori* and between analytic and synthetic, he writes "By this I do not, of course, mean to assign a new sense to these terms, but only to state accurately what earlier writers, Kant in particular, have meant by them." So Frege understands Kant to be using the term "*a priori*" in the sense of Leibniz. As you will see, I do not think that this is entirely right. But, indeed, Leibniz's usage is the traditional one, deriving from discussions of Aristotle's notion of demonstrative knowledge in *Posterior Analytics*, Bk I.2. Demonstrative knowledge is obtained by a syllogistic argument whose ultimate premises are, among other things, "prior." There is nothing in this tradition to suggest that the term "*a priori*" applies primarily to states of knowledge. Though Aristotle speaks of demonstrative knowledge (*episteme*), he is in fact describing demonstrative science and not some species of states of knowing.

But anyway, in defining the notion of *a priori* truth, it is Plato's rather than Aristotle's words that Leibniz (and so Frege) are echoing. In fact, Plato is the first to have introduced the idea of *a priori* truth, at least as far as we know. He first introduces the concept in the *Phaedo* in connection with what he refers to as the "second best method" (99c5–101e1).[8] But his clearest description of this concept occurs in his discussion of *noesis* in the Divided Line simile at the end of Book VI of the *Republic*.[9]

by the other section of the intelligible I mean that which

reason itself lays hold of by the power of dialectic, treating its assumptions not as absolute beginnings but literally as hypothesis, underpinnings, footings, and springboards so to speak, to enable it to rise to that which requires no assumption and is the starting point of all [UP STAGE], and after attaining to that again taking hold of the first dependencies from it, so to proceed downward to a conclusion, making no use whatever of any object of sense but only of pure ideas moving on through ideas to ideas and ending with ideas [DOWN STAGE].

(511b ff.)

It is reasonable to identify the notion of "principles which require no assumptions and are the starting point of all" with Leibniz's notion of primitive truths: they require no proof and have none.

Only Leibniz's assertion that the primitive truths are identical propositions whose opposite contains an explicit contradiction seems to go beyond Plato's conception. But I am not sure that it does and that Leibniz's assertion is not intended merely as an amplification of his statement that the primitive truths do not need proof. For perhaps Leibniz's thinking here is this: the subject of a proposition in geometry, for example, is space – but space as a structure of a certain kind, independent of the question of whether it is our space. This subject, space, then contains the theorems of geometry as its properties. Thus, the theorems of geometry, as assertions about (ideal) space, are analytic. This interpretation of Leibniz would give point to Kant's otherwise somewhat unmotivated emphasis in the Transcendental Aesthetic on the point that space is not a concept. His point would then be that the theorems of geometry are not true of space in the way that a predicate of a proposition may be contained in the subject – since this "containment" is a relation between concepts.

VI

In any case, Leibniz's conception of *a prori* truth may be seen as a part of a tradition in science whose first exponent that we know is Plato. In the exact sciences concepts are employed which do not literally apply to the phenomena and propositions are assumed which are not literally true of them. Thus, in the case of geometry, we do not perceive point, lines, surfaces, etc. and the assumption that, for example, any two magnitudes are comparable is not supported by our actual experience in measuring, in which the results of measuring are inexact and unstable under repeated measurements.[10] It is very reasonable to suppose that realization of this began to dawn on geometers in the fifth century BC, especially after the discovery of incommensurable line segments. From this point of view, Plato's doctrine of Forms may be regarded in part as his description of what had already happened in the case of geometry. Geometry was the study, not of empirical shapes and magnitudes directly, but rather of the kind or form of structure that they imperfectly exemplify. I say "in part," however, because Plato clearly implies in the *Republic* (510d–511b) that geometry is still in the stage of *dianoia* and not *noesis*. Geometers understood that their real subject-matter was not the empirical figures, but the forms that these imperfectly exemplify; but their arguments still had their starting points in the empirical figures. The "up stage" in the quote above expresses Plato's view that, starting with the empirical figures, by a process of dialectic (i.e. analysis), we can arrive at true starting points – Leibniz's primitive truths. In Book VII of the *Republic*, Plato goes on to urge this kind of foundation, not only in arithmetic and geometry, but in astronomy and music theory as well (although in these sciences, it appears, the practitioners had not yet entirely reached the stage of realizing that they were idealizing the phenomena – that the path of a planet is not really a mathematical curve and that the monochord is not really a line segment).

VII

On the conception of apriority I am attributing to Plato and Leibniz, the distinction *a priori*/*a posteriori* certainly does not concern the source of knowledge in any psychological sense having to do with states of knowing. Moreover, it can be misleading to say that it concerns the source of truth in the way that Frege thought that it did. For we have seen that, for him, the same proposition may be a candidate for being either *a priori* or *a posteriori*. It is simply a question of whether it is possible to prove it from primitive truths. This is reflected in Frege's statement that it makes no sense to speak of an *a priori* error (1974: 3). But for Plato and for Leibniz, this could make perfectly good sense. For a candidate for *a priori* truth is about a certain kind of structure and can never be *a posteriori* true, and an empirical statement about the phenomenal world (or, for Leibniz, about the world of simple substances) can never be *a priori*. Thus, Russell's view quoted above that the same proposition can be known both *a priori* and *a posteriori* is mistaken, even leaving aside his belief that this distinction is a psychological one. If A is an empirical proposition, capable of being verified or refuted by empirical evidence, then "A or not-A" – which itself is capable of being empirically verified or refuted – is not *a priori*. The reason is precisely Plato's: since the boundary between what verifies A and what refutes A is blurred, it could happen that neither A nor not-A (or both A and not-A!) obtain.[11]

VIII

As Charles Parsons (1983) notes in "Quine on the philosophy of mathematics," Quine's holistic conception of science does not really even engage the conception of *a priori* truth that we are discussing here. For example, it indeed turns out that there are better theories to describe gravitational attraction than Newton's, which employs a conception of space–time in which space is Euclidean. But this in no way impugns the theorems of Euclidean geometry as truths about Euclidean space. Parsons writes

> The point I wish to make is that what confronts the tribunal of experience is not the pure theory of this type of structure but its being supposed represented by a definite aspect of the actual world. If the resulting theory is abandoned or modified, the form is likely to be that of replacing this structure by a different structure from the mathematician's inventory, even if it was developed only for the purpose.
> (1983: 196)

Quine's monochrome conception of truth does not admit that there are two kinds of truth of geometric propositions, for example: true of Euclidean space and empirically true.

One may feel that the first kind of truth, true of Euclidean space, is trivial and so Quine is right to ignore it, since it simply amounts to provability from the axioms of Euclidean geometry. He speaks in this connection of an "uninterpreted theory-form" and writes that "it can be worthy of study for its structure without its talking about anything" (1960: §52). For Quine, such an investigation takes place within the framework of first-order predicate logic. It is indeed so that "true of Euclidean space" means being provable from the axioms – say Hilbert's axioms in *Foundations of Geometry* (with a more satisfactory statement of the Axioms of Continuity) – of Euclidean geometry. But these axioms are formulated only in second-order logic, which cannot itself be completely axiomatized; and questions concerning Euclidean space remain undecided on the basis of the axioms we now admit. For example, the question of whether there are uncountable sets of points in Euclidean space which are not of the power of the continuum cannot be answered on the basis of primitive truths that we now accept; but, if we ever determine an answer to this question, it will be a truth about Euclidean space and will be *a priori* true. For Quine, the second-order concept of a set of points would itself have to be regarded as part of the uninterpreted "theory-form" and so, from his point of view, Hilbert's axiom system is incomplete and, indeed, essentially incompletable. Of course, Euclidean geometry has an interpretation for Quine, since in the context in which he is speaking he is admitting reference to real numbers (because he believes such reference necessary for doing physical science) and so can recognize the concrete Euclidean space E_3 whose points are triples of real numbers. With the further reduction of the concept of real number to that of a set of rationals – and so on – the essential incompleteness just mentioned is a consequence of the essential incompleteness of Zermelo–Fraenkel set theory. So here is certainly a structure, the cumulative hierarchy of sets, whose truths are surely *a priori* and cannot be understood in terms of deducibility from a given set of axioms. This point will not convince Quine, of course, and in fact will fail to convince others with less idiosyncratic conceptions of mathematics than Quine's. But it will convince those who, like me, are convinced that the conception of set in Zermelo's "Über Grenzzahlen und Mengenbereiche" (1930) does carry with it a conception of truth which goes beyond any axiomatization of it.

So, anyway, I at least do not see that the conception of *a priori* truth, as conceived by Plato and Leibniz, can be reduced to the idea of deduction from axioms (the Down Stage of *noesis*). There is an Up Stage too, which, I believe, is neither arbitrary nor, in the case just mentioned, completable. Naturally, I am quite content to agree that the nonarbitrariness has a naturalistic base, consisting in our agreement in the search for new primitive truths. I am only arguing that there is such agreement, not already secured by a set of axioms. Indeed, it is a consequence of Wittgenstein's analysis in *Philosophical Investigations* of rule-following that agreement secured by a set of axioms already pressuposes a widespread agreement in what we do with them, which ultimately is not secured by axioms or rules.

IX

In spite of Frege's indication to the contrary, Kant's introduction of the concept of apriority by reference to independence of all experience would seem to break from the tradition *at least* to the extent that it makes no reference to proof from primitive truths. Of course, it is a consequence of the traditional concept of *a priori* truth that, if a proposition is *a priori* true, then its truth can be established without reference to the empirical world. But that is a *consequence* and not the definition. There are two difficulties with the traditional definition for Kant. First, unlike either Leibniz or Frege, Kant has no single notion of proof. He distinguishes "discursive proofs," which are "conducted by the agency of words alone (the object in thought)," from "demonstrations," "which, as the term itself indicates, proceed in and through the intuition of the object" (B763). Second, the synthetic *a priori* truths of philosophy have neither discursive proofs nor demonstrations, but only "deductions," that is, transcendental deductions (B761). Indeed, this is the form of Kant's "proofs" of the Principles of Pure Understanding (B202–62). In particular, in the case of the synthetic principles of the understanding, there are no "axioms," that is, general synthetic *a priori* truths, which can serve as Leibniz's primary truths of reason. As I will argue later, Kant was ultimately led to his conception of reasoning in pure mathematics by his desire to establish that applied mathematics is *a priori*. (In fact, to limit the length of the discussion somewhat, I will consider only the case of geometry.) So, both difficulties with the traditional notion of apriority for Kant arise from a rather significant break with the tradition: the fact that he wishes to include under the *a priori* truths propositions about the physical world.

X

Like Kant after him, Leibniz held that the truths of arithmetic and geometry are *a priori*. But, because of his apparent equating of apriority with logical necessity, he seems to be making a much stronger claim than Kant and an altogether unreasonable one. In connection with geometry, however, there is one sense in which Leibniz's claim is weaker than Kant's. Namely, the theorems of geometry are not *a priori* true of physical space for Leibniz, whereas they are for Kant. For Leibniz, points in space are possible points of view of monads; but it is only a contingent truth that any monad exists, since it depends on the condition that God, knowing the best possible world, will create it (and that the best possible world will involve Euclidean space).[12] Space is ideal, and it is as the science of ideal space that geometry is *a priori* true. For example, that between any two points there is a third is for Leibniz a necessary truth. But that between any two monads there is a third can only be contingent for him.

The view that I am attributing to Leibniz seems to be the one most consistent with his position. But one must note that at *Theodicy* §351, he writes

> But with the dimensions of matter it is not thus: the ternary number is determined for it not by the reason of the best, but by a geometrical necessity, because the geometricians have been able to prove that there are only three straight lines perpendicular to one another which can intersect at one and the same point.

Certainly, one might read this as asserting the necessity of applied geometric truth. But there is another reading, based on a *de dicto/de re* ambiguity: let S be in fact the space of points of view of existing monads – call it physical space. For Leibniz, S is in fact Euclidean space. So truths about S are the necessary truths of Euclidean geometry for Leibniz. But when one is speaking about modalities, "truth about S" and "truth about physical space" should be distinguished, since it is a contingent fact that S is physical space.

Of course, the physical space that is Euclidean for Leibniz is the space of monads and its structure is only imperfectly manifested in the phenomenena. The world that appears to our senses is the result of the confusion of perceptions caused by the multitude of monads making up both the phenomenon and our sense organs. For Kant, on the other hand, geometry is not only *a priori* true of the physical world, but the physical world *is* the phenomenal world. The relation of Leibniz and Kant to Plato in this connection is interesting. Like Kant and unlike Leibniz, Plato does not seem to have had a conception of space which is a model of geometry. The space of the *Timaeus* is the locus of sensible things, not of points, lines, etc. (see 52b-c). But like Leibniz, Plato believed that geometry applies only imperfectly to the phenomena. On the other hand, from his conception of ideal space as a model of geometry, Leibniz provides an account of both why geometry applies at all to the phenomena and why it applies imperfectly in terms of confused perceptions arising from the sheer complexity of the aggregate of the perceptions of individual monads, which themselves determine the spatial relationships in ideal space.

XI

Leibniz certainly does explicitly equate the apriority of geometry with logical necessity. Indeed, for him, all truths that do not imply the existence of a simple substance are logically necessary. Although I made a suggestion concerning this above, at the end of section V, it is not really clear why he thought it appropriate to refer to the primary truths about ideal structures – for example, the primary truths of geometry – as identities; but, by doing so, he obscures our distinction between purely logical truth and truth about ideal structures. As Kant is often read, it was he who made the appropriate distinction: namely it is his distinction between analytic and synthetic *a priori* truth. However, I think that there is good reason to question the identification of his notion of analytic truth with our notion of logical truth. But I will return to this below.

XII

Kant's criticism of Leibniz goes beyond the subsumption of the truths of geometry and arithmetic under logical truths. He attributes this subsumption to the attempt to treat what belongs to Sensibility as though it belonged to the Understanding. Thus, in place of Leibniz's conception of geometry as the science of ideal

THE CONCEPT OF *A PRIORI* TRUTH

space he conceives of it as the science of space as the pure form of sensible intuition.[13]

In this respect, Kant has often been accused of obliterating another distinction, namely that between pure and applied geometry. For Leibniz, this is the distinction between geometry as the *a priori* science of ideal space and geometry as the *a posteriori* science of physical space (and, of course, he further distinguishes both of these from the inexact science of phenomenal space). But, as Michael Friedman (1985) has pointed out, Kant also maintains a distinction between pure and applied geometry, although the distinction takes a different form: pure geometry is the science of space as pure form of sensible intuition and applied geometry is the science of phenomenal space. That Kant regards these as essentially distinct is clear from his discussion at B206–7 of the Principle of the Axioms of Intuition, namely that all intuitions are extensive magnitudes. Thus:

> What geometry asserts of pure intuition is therefore undeniably valid of empirical intuition. The idle objections, that objects of the senses may not conform to such rules of construction in space as that of the infinite divisibility of lines or angles, must be given up. For if these objections hold good, we deny the objective validity of space....

Note that he is saying that, if geometry is not valid concerning sensible objects, then *space* is not objectively valid – in other words, the pure form of sensible intuition is not in fact the form exemplified by appearances. So clearly he distinguishes between geometry as the science of the pure form of sensible intuition (pure geometry) and geometry as the empirical science of appearances (applied geometry).

There is a source of confusion in the term "pure form of sensible intuition" in that, if space is in fact the form of sensible intuition, then what is true of space would seem to be necessarily true of appearance regarding their geometric form. But I think that we have another *de dicto/de re* ambiguity here. At B160, Kant writes

> But space and time are represented *a priori* not merely as *forms* of sensible intuition, but as themselves *intuitions* which contain a manifold.

In the footnote to this passage, Kant contrasts space as the *form of intuition* with space as a *formal intuition*. As formal intuition,

47

space is the subject-matter of pure geometry; as the form of sensible intuition, it is the subject of applied geometry. In the Transcendental Exposition of Space, Kant argues from the fact that geometry is the science of space and is founded on construction in pure intuition that space is the pure form of sensible intuition. But we must understand him here as thinking of space as formal intuition, since the objective validity of geometry is established only in the Principles of Pure Understanding. (It is true that at B119, Kant writes that we "have already, by means of a transcendental deduction, traced the concepts of space and time to their sources, and have explained and determined their *a priori* objective validity." The "transcendental deduction" in question is presumably the transcendental exposition of space and time. I can only suggest that, in the case of space, "objective validity" in this passage refers to the validity of geometry as the science of space *qua* formal intuition. That is surely the sense of the expression in which Kant is attempting to establish objective validity in the Transcendental Exposition.) The main point, in any case, of the distinction between space as form of intuition and space as formal intuition is that we can study the structure of space independently of the question of whether it is in fact the structure of (our) space.

So both Leibniz and Kant maintain a distinction between pure and applied geometry. Pure geometry is concerned with ideal space and with space as formal intuition, respectively, and applied geometry is concerned with the spatial relationships of actual monads and of appearances, respectively.

XIII

When Kant accuses Leibniz of assigning to Understanding, the realm of concepts, what properly belongs to Sensibility, the realm of intuitions, however, his point is not entirely well taken. Since he is here attacking Leibniz's view that geometric truth is analytic, the issue concerns the nature of pure, not applied, geometry. Admittedly, he is associating Leibniz with Wolff in his accusation and, for Wolff, the distinction between pure and applied geometry (in Leibniz's sense) is obliterated. So, without knowing Wolff's position, it is perhaps not possible to fully understand the force of Kant's criticism. But it is true that, for Kant, the subject-matter of pure geometry is not sensible intuition but its pure form, space. And the fact that sensible intuition involves a nonconceptual

THE CONCEPT OF *A PRIORI* TRUTH

element does not imply that its pure form does as well. Kant argued that space is not a concept; but that is not the issue. The question is: why did Kant believe that our knowledge of space (as pure form) is not entirely conceptual? There seem to be two directions in which to go in attempting to answer this question.

One direction, taken by Bertrand Russell, for example in *The Principles of Mathematics* (1903), and by Michael Friedman (1985), is to attribute Kant's theory of space to his narrow conception of logical form. Certainly his official logical doctrines seem to restrict concepts to unary (nonrelational) concepts, the logical form of propositions to propositional combinations of universal, particular or singular propositions and logical inference to inferences which can be formalized in monadic predicate logic. On this conception, the result of abstraction of sense experience from all concepts still retains a "form," since it still involves spatial relationships, for example. Moreover, Kant's distinction between discursive proof and demonstration can be understood on this view: the former is contained in monadic predicate logic and so does not include all geometric reasoning. The inference, for example, from the statement

(a)   Between any two points there is always a third.

to the statement that between any two points there are always indefinitely many cannot be regarded as a logical inference for Kant on this view, since he did not recognize the relevant logical forms of the statements.

Concerning the latter inference, one should distinguish two questions. One concerns the inference, for any particular number $n > 1$, from (a) to

(b)   Between any two points there are n points.

The other question concerns the inference from (a) to

(c)   Between any two points there are infinitely many points.

Friedman seems to believe that Kant would not regard the first of these inferences to be a logical inference. He writes that "for Kant, one cannot represent or capture the idea of infinity formally or conceptually: one cannot represent the infinity of points on a line by a formal theory like [the theory of dense linear order]" (1985: 466). His point is not that, for Kant, one cannot have a concept of infinitely many points but that no set of axioms is such that all of its models must be infinite on Kant's conception of logic. Thus, it is the inference from (a) to (b) (for arbitrary n) that is in question.

The question of whether the inference from (a) to (c) is a logical inference, even after we add to the premise the relevant other axioms for the betweenness relation, is more complicated. It becomes a logical inference only if we admit higher-order logic and, in particular, the definition of the class of finite sets of points as the least class (i.e. intersection of all classes) containing the null set and, for every point A, containing $S \cup \{A\}$ whenever it contains S.

It is interesting that this definition of finite set derives from the treatment of the thory of formal sequences in Frege's *Begriffschrift* and in Dedekind's *Was sind und was sollen die Zahlen?*, which was the basis of Frege's attempt to refute Kant's view that arithmetic is synthetic. One must certainly agree with Friedman that Kant's view that arithmetic is synthetic is closely connected with the lack of a purely logical treatment of iteration (or formal sequences). But notice also that the question of whether or not there is such a purely logical treatment does not depend upon whether one restricts logic to monadic logic or not. Rather it depends upon whether one admits higher-order logic. And Kant's thesis that geometry is synthetic is certainly not a consequence of his restriction of logic to first-order logic.

One obvious response to the Russell-Friedman interpretation is that Frege, whose conception of logic extended to higher-order quantification theory, nevertheless defended Kant's conception of geometry. But, of course, that does not prove Russell and Friedman to be wrong about Kant, himself.

Frege also possessed something else, at least in his later writings supporting Kant's view of geometry, that Kant did not, namely an adequate axiom system for Euclidean geometry, in particular in Hilbert's *Foundations of Geometry* (1899), and a proof of the categoricity of these axioms. Since it was also known how to build the model E_3 (see Section VII above) and since Frege thought that the theory of real numbers was analytic, there is a sense in which he – as opposed to Kant – might have regarded geometry as analytic. Namely, any truth in the geometry of "real" space is at the same time (on account of categoricity) a truth about E_3 and these truths are analytic. Of course, Frege would have objected to this line on the grounds that the truth about real space is not the same proposition as the corresponding truth about E_3. But why not? Why should not Frege have regarded the notion of truth about the pure form of intuition as being well analyzed as truth concerning E_3? In Kant's time, not even the basic properties of order on the line were axiomatized. Thus, he might well have doubted the adequacy of the axiomatic conception of geometry without this being a consequence of his theory of logic. Even on our conception of logic, the axiomatic foundation of geometry of his time is inadequate. Nor did the movement towards a rigorous axiomatic foundation for geometry either arise from or depend on the development of the science of logic. It is plausible to suppose that the study of non-Euclidean and projective geometries, because of their formal character, led to a clearer conception of purely logical reasoning from assumptions (axioms) and enhanced the ideal of a rigorous axiomatic development of geometry. But neither Hilbert's book nor, certainly, Pasch's *Lectures on Modern Geometry* (1882), in which an adequate axiomatization of the ordering of the points on the line was first given, betray any influence at all of the contemporary developments in logic. In short, the question of belief in the possibility of an adequate axiomatic development of geometry seems independent of the development of the formal logic of relations.

Of course, there is a sense in which, no matter what conception of logic one has, one can regard any science as a deductive science: find a logically independent subset of the set of all true propositions of the science from which all the other propositions of the science can be deduced and call the members of this subset "axioms." (Providing the conception of logic in question includes classical propositional logic, this can always be done.) But the real issue in connection with the Russell–Friedman interpretation is whether the traditional development of geometry as an axiomatic theory – admitting that there are gaps in some of the proofs – was understood by Kant in terms of purely logical deduction of propositions from a set of axioms.

I am inclined to agree with Henry Allison in *Kant's Transcendental Idealism: An Interpretation and Defense* (1983: 73–4) that Kant's conception of analyticity is not accurately reflected in his first characterization in the *Critique of Pure Reason*, namely that the predicate is contained in the subject (B10) – at least, not in the sense in which we normally read that formula. On the other hand, Allison contends that Kant's real notion of analyticity is given later, at B190-1, in terms of the Principle of Noncontradiction. But, in fact, Kant's statement of that principle, "no predicate contradictory of a thing can belong to it," is, on the face of it, hardly more adequate. Indeed, as usually understood, neither characterization of analytic truth would seem to support Kant's assertion at B204 of the analyticity of the proposition that, if equals be added or subtracted from equals, the results are equal. Kant's argument for analyticity, namely that "I am immediately conscious of the identity of the production of the one magnitude with the production of the other," cannot literally be based on either of the characterizations of analyticity, as they are usually understood. In fact, as usually understood, both characterizations apply only to propositions of the form "All S are (are not) P." Leibniz, too, characterizes logical truth as meaning that the predicate is contained in the subject. But, on p. 479 of the *New Essays* (with which Kant was acquainted), he explicitly admits logical inferences which are not syllogistic and, indeed, are not formalizable in monadic logic, though they are logically valid inferences in our sense. Of course, as we have already noted, Leibniz characterizes as logical truths the truths of Euclidean geometry.

In view of these examples and of Leibniz's repeated claims that space is relational and that the truths of geometry are logically true, I think that it is truly surprising that Russell, in his book on Leibniz, attributed the latter's whole philosophy to his belief that all propositions are of subject–predicate form. It is true that, for Leibniz, no proposition can express a relation between two or more *substances*. But that follows from his metaphysical conception of substance and the same is by no means true, say, of points (possible points of view of monads). Kant understood Leibniz better than Russell did on this point:

> For how are we to think it to be possible, when several substances exist, that, from the existence of one, something (as effect) can follow in regard to the existence of the others.... Leibniz, in attributing to the substances of the world, as thought through the understanding alone, a community, had therefore to resort to the mediating intervention of a Deity.
>
> (B292-3)

However one is to understand Kant's definition(s) of analyticity,

THE CONCEPT OF *A PRIORI* TRUTH

there seems to me to be no real evidence that he understood it in a sense narrower than first-order logical validity. The question is, confronted with an inference formalizable in first order, would he have rejected it as a logical inference? Kant nowhere explicitly questions the analyticity of such an inference. His argument against the dogmatists and for his conception of geometry is that, for example, the theorem that the interior angles of a triangle equal two right angles is not a logical truth – and he is absolutely correct on the most enlightened conception of logic.

Indeed, much less having a narrower conception of analyticity than our conception of logic, I think that it is, in some respects, wider. Certainly that would seem to be indicated by the above example, "Equals added or subtracted from equals are equal," of an analytic proposition. It is unclear to me how Kant would understand this example as a subject–predicate proposition; but it is analytic presumably because it follows from the meaning of equality and sum or difference. But, in the same way, it seems reasonable to regard purely universal propositions about the betweenness relation on a line – such as that, for any three distinct points on the line, precisely one of them lies between the other two – as analytic because they follow from the meaning of "line" and "betweenness." Unlike (a), they imply no existence statements. Moreover, there seems to be no reason to doubt that, confronted with the implication from (a) to (b) (for any particular n), Kant would have regarded it to be analytic. The properties of betweenness required are

(d) If R is between P and Q, then P, Q, and R are distinct points.

(e) If R is between P and Q and S is between P and R, then S is between P and Q.

These again are purely universal propositions which could be taken to follow from the meaning of betweenness. Of course, since Kant nowhere discusses the properties of the betweenness relation and the axioms for this relation were not available to him, the question is not whether he explicitly regarded the above propositions as analytic but whether his notion of analyticity, in its intension, extends to such propositions. Whatever logical truth meant for Leibniz or for Kant – however they understood the Principle of Noncontradiction and the formula "the predicate is contained in the subject" – I think that both of them gave it a wider meaning than it has for us. "Body is extended" is analytic for Kant and, I expect, "All bachelors are unmarried" would be analytic for both Leibniz and Kant. Thus, I do not think that Kant's complaint against Leibniz is exactly ours. Some of the (for us) nonlogical truths that Leibniz considered to be logical truths Kant also counted as analytic. In both cases, their conception of logical truth was too wide: too much was packed into the idea of the predicate being contained in the subject.

We have noted that the Common Notions, such as "Equals added to (or subtracted from) equals are equal," are analytic for Kant. It is very difficult to square this fact with the Russell-Friedman reading, since "equals" is a relation – and *not* the trivial relation of identity. It is a relation which holds, for example, between any two triangles with the same base and height. Friedman recognizes this but, nevertheless, believes that Kant would reject the inference from (a) to (b) as a logical inference. But his argument *here* is not that the premise and conclusion involve a relation, but that recognizing the validity of the inference depends upon recognizing the quantifier form "For all x and y, there exists a z such that" of (a). Of course, this is not a matter of restricting logic to monadic logic. For example, in monadic logic one cannot express the Common Notion "Equals added to equals are equal" and one cannot express and derive the inference from "All horses are animals" to "All heads of horses are heads of animals," although neither of these involve the "For all, there exists" quantifier form. Of course, the inference from (a) to (b) is not a logical inference in our sense; but Friedman's reading would have Kant reject even the inference from (a), (d), and (e) to (b) (with $n = 2$, say) as a logical inference. I would prefer to have a reading which explains Kant's particular philosophy of geometry without attributing to him falsehoods which he himself nowhere expresses.

XIV

Friedman (1985) notes another source of Kant's conception of pure intuition that is more to the point than his theory of logic, namely the central role for him of the notion of construction in geometry. For Kant, geometric proof essentially involves construction. The above theorem, on triangles, for example, must begin with the construction of the concept "triangle" and then of some auxiliary lines (B744). At B741–2 Kant describes what he means by constructing a concept:

> To *construct* a concept means to exhibit *a priori* the intuition which corresponds to the concept. For the construction of a concept we therefore need a *nonempirical* intuition. The latter must, as intuition, be a *single* object, and yet none the less, as the construction of a concept (a universal representation), it must in its representation express universal validity for all possible intuitions which fall under the same concept. Thus I construct a triangle by representing the object which corresponds to this concept either by imagination alone, in pure intuition, or in accordance therewith also on paper, in empirical intuition – in both cases completely *a priori*, without having borrowed the pattern from any experience. The single figure which we draw is empirical, and yet it serves to express the concept, without impairing its universality. For in this empirical intuition we consider only the act whereby we construct the concept, and abstract from the many determinations (for instance, the magnitude of the sides and of the angles), which are quite indifferent, as not altering the concept "triangle".

For the construction of the concept "triangle" we need a nonempirical intuition. As an intuition, it is a single object but it must also "in its representation" express universal validity for all possible triangles. Kant often writes that the only objects are empirical objects, that is, sensible intuitions. But he is not consistent about that. Indeed, the following passage contains both uses of the term:

> It does, indeed, seem as if the possibility of a triangle could be known from its concept in itself (since it is certainly independent of experience); for we can in fact give it an object completely *a priori*, that is, can construct it. But since this is only the form of an object, it would remain a mere product of the imagination, and the possibility of its object would still be uncertain.
>
> (B271)

The conflict in terminology occurs because Kant wishes to regard a form of an object (i.e. of a sensible intuition) also as a formal intuition, a nonempirical object. But note that, whether or not we should call it an object, the pure form of triangles, that is, what we construct in pure intuition when we construct the concept "triangle," is not itself a triangle. If it were, then it would have to be right-angled, obtuse-angled, or acute-angled and so would not be the form of triangles of the other two kinds. So, as Friedman has argued in his paper "Kant on concepts and intuitions in the mathematical sciences" (forthcoming), pure intuition does not provide a model for Euclidean geometry; in fact, it does not even provide geometric objects: line segments, triangles, circles, and the like.

Jaakko Hintikka (1973) had an interesting idea for understanding the generic character of "objects" constructed in pure intuition. Namely, he understands it in terms of the formal logical inference of Existential Quantifier Elimination: from "There is an x such that $A(x)$" and "If $A(b)$, then B," where $A(x)$ and B do not contain the new individual constant b, infer B. Constructing in pure intuition is to be understood as passing from "There is an x such that $A(x)$" to "$A(b)$" (with the purpose of then deducing "B"). Thus, if "$A(x)$" expresses that x is a triangle, then b stands for the generic triangle. But this idea will not work, since it does not account for why "There is an x such that $A(x)$" is true in applied geometry. We will see that Kant's argument for the apriority of applied geometry (such as it is) depends on the premise that the *act* of construction of a figure in pure intuition is precisely what is involved in the synthesis of empirical figures. Hintikka's formal account of construction does not admit this structure of a construction as a (temporal) act.

XV

On the other hand, Friedman draws from the fact that pure intuition does not provide a model of geometry a further consequence which seems more questionable. Namely, he attacks a standard account of Kant's conception of geometry according to which Kant regards geometry as a deductive system based on definitions, postulates, and axioms, where the inferences involved are themselves analytic. According to this account, the role of construction and pure intuition lies in the fact that the axioms and postulates are justified by construction in pure intuition. This account is the other and, to me, more plausible direction to take in accounting for Kant's view that geometric reasoning is not discursive. It certainly has textual support. Kant writes:

THE CONCEPT OF *A PRIORI* TRUTH

For as it was found that all mathematical inferences proceed in accordance with the principle of contradiction (which the nature of all apodeictic certainty requires), it was supposed that the fundamental propositions could also be recognized from that principle. This is erroneous. For a synthetic proposition can indeed be comprehended in accordance with the principle of contradiction, but only if another synthetic proposition is presupposed from which it can be derived, and never in itself.

(B14)

The antecedent clause of the first sentence would seem to stipulate that mathematical inference is logical inference. Friedman (1985: 489–90) questions this on the grounds that it need not mean that *all* mathematical inferences are logical, only that some are. But then surely the force of the connection between the first and second clause would be lost. There are two kinds of "fundamental propositions," that is, synthetic *a priori* propositions of geometry, that Kant explicitly mentions in the *Critique of Pure Reason*, namely postulates and axioms. Concerning the postulates, he writes:

Now in mathematics a postulate means a practical proposition which contains nothing save the synthesis through which we first give ourselves an object and generate its concept – for instance, with a given line, to describe a circle on a plane from a given point. Such a proposition cannot be proved, since the procedure which it demands is exactly that through which we first generate the concept of such a figure.

(B287)

Concerning the axioms, he writes:

The mathematics of space (geometry) is based upon this successive synthesis of the productive imagination in the generation of figures. This is the basis of the axioms which formulate the conditions of sensible *a priori* intuition under which alone the schema of a pure concept of outer appearance can arise – for instance, that between two points only one straight line is possible, or that two straight lines cannot enclose a space, etc.

(B204)

I think that these passages strongly support the reading of Kant that Friedman opposes. This reading does not require that pure intuition provide a model of Euclidean geometry, in the sense that there be "objects of pure intuition," namely points, lines, and planes. Indeed, the second and last of the quoted passages make it clear that this is not the sense in which the postulates and axioms are founded in pure intuition.

XVI

Kant's discussion of mathematical postulates applies to Euclid's "To construct" postulates, namely Postulates 1–3 and 5, but not to Postulate 4, which asserts that all the right angles are equal. Construction is only part of geometric proof.[14] In the typical cut-and-paste proofs that two figures are equal, such as the proof that the interior angles of a triangle are equal to two right angles, construction applies to reduce the equation, by means of the Common Notions (equals added or subtracted from equals are equal), to a set of equations which have already been established. Kant takes the Common Notions to be analytic. But such arguments must begin with *some* nontrivial equations, such as Postulate 4 or the congruence axioms of Hilbert. The fact is that he does not seem to have considered such propositions: they are not postulates in his sense, nor are they axioms. For, as I read his above description of the notion of an axiom, as well as all but one of his examples, an axiom of geometry states the sufficient conditions for the determination of a geometric object. Thus, two distinct points on a line determine the line, the radius and center of a circle determine it and three noncoplanar lines determine space (i.e. space is three dimensional). The one example that does not quite fit this analysis is one cited by Kant in the passage above, that two lines do not enclose a space. But this proposition ought to mean that, given two lines K and L in the plane P and a point A in P on neither of them, there is a line in P through A which has at most one point which is on either K or L. And this follows from two propositions: (i) through any point in P not on a line L in P there is a line in P parallel to L and (ii) there is one and only one line containing two distinct given points. But these are consequences of postulates and axioms in the sense of these terms that I attribute to Kant. So I am inclined to think that Kant was simply being careless here in calling the proposition in question an axiom. If this is right, then the axioms are simply concerned with the uniqueness of figures given by some construction, not with the equality of distinct figures.[15] But then Kant seems to have left completely out of account in his theory of geometry the status of the basic equations – for example, by means of which one would prove that the equality of two sides and the included angle implies the equality of the other corresponding parts of two triangles. In his published works, at least, the only discussion of proofs of congruence is in *Prolegomena* §12, where he merely writes:

> All proofs of the complete congruence of two figures (when one can be replaced at all points by the other) finally come to this, that they coincide with each other; which is obviously nothing other than a synthetic proposition resting on immediate intuition. This intuition must be given pure and *a priori*, otherwise the proposition could not hold as apodictically certain, but would have only empirical certainty.

Well thank you, Mr Kant. Presumably he does not mean that all congruence theorems rest on immediate intuition, except in the sense that they derive ultimately from basic equations or congruence propositions which rest on immediate intuition.[16] But these basic equations or congruence propositions are neither postulates nor axioms, as I understand Kant to be using these terms, nor does he attempt to explain *how* they "rest on immediate intuition," as he attempts to do in the case of postulates and axioms.

XVII

Kant's emphasis on construction is close to the spirit of Euclid. Many of the latter's propositions are to the effect that a figure with a certain property can be constructed. Indeed, the propositions are formulated as problems: "To construct a figure x such that $P(x)$." And the proofs, allowing for some gaps in Euclid's argument, consist of iterated applications of the "To construct" postulates. It is possible – and I think entirely reasonable – to interpret these "To construct an x such that $P(x)$" propositions as having a meaning that is stronger than is obtained by translating them into the corresponding existence proposition: "There exists an x such that $P(x)$." Namely, one might interpret them as meaning that the figure in question is obtained by iterating Euclidean constructions (from a given collection of points) – in other words, that the figure is Euclidean constructible. Friedman (1985: 466) suggests that, in fact, Kant understood existence theorems in geometry in this way and, for that reason, could not have adopted the conception of geometric proof as discursive proof from axioms. But notice that, if this was Kant's problem, it does not arise from the view that logic is monadic but rather, as in the case of the inference from (a) to (c), from the view that it is first order. For it is only in higher-order logic that we can express the notion of a figure being Euclidean constructible – namely that it be in the least class (i.e. the intersection of all classes) of figures closed under ruler and compass constructions. But, in any case, Friedman's idea is that the act of construction is essential to Kant's conception of geometric proof because, for him, a statement such as that there are indefinitely many points on a line can only be understood as an assertion that we possess a means of actually constructing them.[17]

There is some question as to whether diagrams, either sensible or in pure intuition, were essential in proof for Euclid. There is strong internal evidence that they are not.[18] Certainly Leibniz is very explicit on this point in *New Essays*:

> You must understand that geometers do not derive their proofs from diagrams, although the expository approach makes it seem so. The cogency of the demonstration is independent of the diagram, whose only role is to make it easier to understand what is meant and to fix one's attention. It is universal propositions, i.e. definitions and axioms and theorems which have already been demonstrated, that make up the reasoning, and they would sustain it even if there were no diagram.
>
> (1981: 360–1)

Of course, Kant was contesting Leibniz's conception of geometric reasoning. But the question is what aspect of that conception was he contesting. Explicitly, it was Leibniz's view that the primitive truths of geometry are "identities," that is, follow from the Principle of Noncontradiction. He nowhere criticizes Leibniz because logical inference cannot account for the particular content of "To construct" theorems. But Leibniz's position is anyway interesting because it indicates that the view that diagrams are essential to geometric proof is not part of the tradition that Kant inherited. So perhaps, rather than attempting to account for his conception of pure intuition in terms of the need to find the appropriate medium

THE CONCEPT OF *A PRIORI* TRUTH

of construction, one should ask for his motives for giving to construction the essential role that it plays in his conception of geometry.

XVIII

As I have indicated above, I believe that the answer to this question lies in Kant's desire to establish that the theorems of geometry are *a priori* true as propositions of *applied* geometry. Quite simply, his argument that the pure-form-of-sensible-intuition is in fact the form of sensible intuition is that the act by which we apprehend an appearance as, say, a triangle is precisely the same as the act by which we construct the concept "triangle" in pure intuition. Thus, in Kant's "proof" of the "principle of the axioms of intuition," namely that all intuitions are extensive magnitudes, he writes:

> Appearances, in their formal aspect, contain an intuition in space and time, which conditions them, one and all, *a priori*. They cannot be apprehended, that is, taken up into empirical consciousness, save through that synthesis of the manifold whereby the representations of a determinate space or time are generated, that is, through the combination of the homogenous manifold and consciousness of its synthetic unity.
>
> (B202-3)

To connect this synthesis with the notion of construction, recall the passage quoted above from B204: "The mathematics of space (geometry) is based on this successive synthesis of the productive imagination in the generation of figures."

Perhaps one reason why Kant scholars might wish to find other motivations for his philosophy of geometry than this argument for the apriority of applied geometry is that it is such a bad argument. Kant nowhere seems to be aware of the fact that the assertion that space is Euclidean is an assertion in part about the behavior of measuring devices – transported rigid bodies, light rays, etc. The objects to which we would apply Euclidean geometry are not simply "figures" which we generate by a "successive synthesis of the productive imagination"; but, most importantly, geometric proof does not consist simply in "generation" or construction. It also involves deriving equations from basic equations such as the equality of all right triangles; and Kant's proof that all intuitions are extensive magnitudes does not touch on this.

But Kant's reputation is not enhanced by rejecting this argument as the point of his particular philosophy of geometry and attributing the latter instead to a conception of logic which was restricted even for his day. For, after all, Kant *did* publish this argument anyway and so why attribute to him, quite unnecessarily, another mistake as well?

NOTES

*I read versions of this essay at the Boston Colloquium for the Philosophy of Science in November 1988 and at the University of Illinois in Chicago in March 1989. I profited from subsequent discussions in both cases and, especially, from the prepared comments of George Boolos in Boston. But, if I am happier with this, the third, version, it is probably because I have managed to cram in most of what I wanted to say and because I have given up thinking of it as a coherent essay.

1. See Quine's "Truth by convention" (1936), reprinted in the collection of his essays *The Ways of Paradox* (1966). In particular, see p. 95 of the latter book.
2. On p. 24 Russell (1937) writes that the concept of apriority concerns the source of knowledge as opposed to the source of truth. On p. 160 he assigns this question to psychology.
3. In fact, what he writes is that the distinction between the empirical and the *a priori* seems to depend on this confusion (p. 24); but immediately above this passage he indicates that he is using "empirical" to mean contingent.
4. For example, the components A and B of the judgement "if A, then B" or "A or B" are problematic judgements.
5. Russell disputes this (pp. 32–3) on the basis of passages from around 1686 in which Leibniz suggests that every true proposition has an *a priori* proof. In the case of contingent propositions, the "proof" may involve the consideration of all possible worlds and so be infinite and not available to us. And even this proof does not establish truth independent of the fact that God, knowing what is best, will ordain it. In later writings, however, Leibniz is quite explicit that apriority coincides with necessity. In any case, the passages in question are far from supporting Russell's conclusion that apriority is a psychological notion for Leibniz. For they are still characterizing *a priori* propositions in terms of the source of their truth, not in terms of our knowledge of them. (Indeed, if the proof is infinite, presumably we would not "know it *a priori.*")
6. Of course, in the 1790 remark, Kant may only be noting that the notion of apriority arose in logic without endorsing the view that it is a logical notion. That geometry applies to the empirical world is a transcendental truth for Kant.
7. Frege also seems to go beyond Leibniz's notion of *a priori* in that he admits as *a priori* the "general proposition that the inductive method

can establish the truth of a law, or at least some probability for it" (1884: 4, footnote), which is not necessary in Leibniz's sense. Other than this, the only *a priori* truths Frege mentions are analytic truths and the synthetic truths of geometry.

8   This passage is discussed in my paper, "Plato's second best method" (1986b).
9   Shorey's translation.
10 See *Phaedo*, 74b.
11 The rationale for Leibniz in the case of propositions about the world of substance is somewhat different: "A or not-A" may have existential import and so not be true. For example, A may contain a name of some substance.
12 Why, for Leibniz, should this not have been true of any science? Namely, every such science has an *a priori* part, dealing with the possibilities of substance, and the contingent part, which adds the claim that the possibilities are realized.
13 I do not mean to suggest that Kant entirely understood Leibniz's conception of space. For example in the *Inaugural Dissertation* §5D he suggests that Leibniz held that space is "a relation of existent things, vanishing therefore if things be annihilated, and not thinkable except in actual things...."
14 Thus, Friedman's remark (1985: 457) that a geometric proof for Kant is a spatio-temporal object is not strictly true, although it is true that such a proof essentially involves a spatio-temporal object.
15 Interpretations of Kant which take the criterion for whether or not a proposition is synthetic to be whether or not it has existential import seem to overlook Kant's notion of an axiom and the fact that the axioms are synthetic. This does not quite apply to Hintikka, since on his view, Kant also takes to be synthetic propositions which imply the *nonexistence* of objects, and the axioms may be understood in that way. (For example, "There are not two lines containing a given two points.") On the other hand, Hintikka's interpretation does not account for the syntheticity of Euclid's Postulate 4. Gordan G. Brittan (1978) does take existential import as the criterion of syntheticity for Kant. He (p. 71) seems to misunderstand how Kant is using the term "axiom" and takes him to be part of a tradition which refers to the Common Notions as axioms.
16 "Equations" in the sense of statements of equality of figures *qua* magnitudes (*quanta*) or, as we can say, identity of their magnitudes (*quantitas*). This was the sense of "ισοσ" in Euclid. The notion of congruence does not explicitly occur in his work. The connection between the two notions is that congruent figures are equal, though the converse of course need not be so.
17 On this reading there is, of course, a close connection between Kant's conception and L.E.J. Brouwer's Intuitionism.
18 First of all, there is Euclid's platonic background. But more specifically, there is internal evidence. For example, for many theorems in Euclid, he gives only one diagram where, if the diagram were essential, he should give more than one, for example, depending upon

whether an angle is acute, obtuse, or right. Again, in Book V, Euclid is explicitly treating ratios of all kinds of (like) geometric magnitudes – lines, surfaces, and solids – but his diagrams are only of lines.

REFERENCES

Allison, H. (1983) *Kant's Transcendental Idealism: An Interpretation and Defense*, New Haven, CT, and London: Yale University Press.

Brittan, G.G. (1978) *Kant's Theory of Science*, Princeton, NJ: Princeton University Press.

Frege, G. (1884) *Die Grundlagen der Arithmetik*, Breslau: Koebner. English translation in *The Foundations of Arithmetic*, Evanston, Ill.: Northwestern University Press, 1974.

Friedman, M. (1985) "Kant's theory of geometry," *Philosophical Review* 94, 455–506.

────── (forthcoming) "Kant on concepts and intuitions in the mathematical science," *Synthese*.

Hintikka, J. (1973) *Logic, Language-games and Information: Kantian Themes in the Philosophy of Logic*, Oxford: Clarendon Press.

Kitcher, P. (1983) *The Nature of Mathematical Knowledge*, Oxford: Oxford University Press.

Kripke, S. (1980) *Naming and Necessity*, Cambridge, MA: Harvard University Press.

Leibniz, G.W. (1948) *The Monadology and Other Philosophical Writings* (trans. R. Lotta), London: Oxford University Press.

────── (1951) *Theodicy: Essays on the Goodness of God, the Freedom of Man, and the Origin of Evil* (A. Farrer, ed., trans. E.M. Huggard), London: Routledge & Kegan Paul.

────── (1981) *New Essays on Human Understanding* (P. Remnant and J. Bennett trans and eds), London: Cambridge University Press.

Parsons, C. (1983) "Quine on the philosophy of mathematics," *Mathematics in Philosophy: Selected Essays*, Ithaca, NY: Cornell University Press, 176–205.

Plato (1937) *The Republic*, with an English translation by P. Shorey. Loeb Classical Library vol. 5, 6. Cambridge, MA: Harvard University Press.

Quine, W.V. (1936) "Truth by convention." Reprinted in *The Ways of Paradox*, New York: Random House, 1966.

────── (1960) *Word and Object*, Cambridge, MA: MIT Press.

Russell, B. (1903) *The Principles of Mathematics*, Cambridge: Cambridge University Press.

────── (1937) *A Critical Exposition of the Philosophy of Leibniz*, London: George Allen and Unwin

Tait, W.W. (1986a) "Truth and proof: the platonism of mathematics," *Synthese* 69: 341–70.

────── (1986b) "Plato's second best method," *Review of Metaphysics* 39 (3): 455–82.

Zermelo, E. (1930) "Über Grenzzahlen und Mengenbereiche," *Fundamenta Mathematica*, 16: 29–47.

This work was supported by grant DIR-9011694 from the National Science Foundation.

3
LOGICISM
Steven J. Wagner

SUMMARY

The "first-generation" logicism of Frege and Dedekind differs from subsequent versions by claiming a purely rational status for mathematics. An updated version of this claim is defensible. Mathematics is *a priori* in the sense of needing empirical data only to compensate for human limits of memory and attention. It is analytic in the sense that any rational beings satisfying minimal conditions would have reason to develop forms of arithmetic and set theory. This version of logicism is immune to standard objections. In particular, it avoids Philip Kitcher's critique. My notion of analyticity, however, is more general than its traditional counterparts. One point is that analytic knowledge may be conjectural. Logicism is, ultimately, a thesis about human cognitive capacities and mechanisms.

Everyone knows that logicism is half right. What is right is the identification of mathematics with set theory. Standard mathematics can clearly be done in set theory – even though this claim involves deep problems. And set-theoretic mathematics is a structure of proofs in the logic the logicists invented for their purpose. What is wrong is the epistemology. Logicists aimed to derive mathematics from self-evident truths by sheer logic. But Russell's paradox and Gödel's theorems frustrate this aim, while the new epistemology of Peirce, Quine, Wittgenstein and others shows that the entire project is misconceived. So success and failure divide neatly along disciplinary lines. The mathematics conquered, the philosophy collapsed.

Quine above all has promoted this assessment, developing and clarifying the set-theoretic foundations of mathematics in his logical work while rejecting the old epistemological vision. And up to a point, his judgment must stand. Whatever the future may bring, the reduction of mathematics (as we know it) to set theory (as we know it) must remain a landmark. Yet a return to the epistemologies of Frege, or Russell, or Carnap seems out of the question. But I think that Quine, too, is wrong. I will here reconsider the epistemology of logicism, in particular urging a favorable view of Frege (my principal subject) and Dedekind. Although their logicism failed, I believe some of its main philosophical elements were abandoned too quickly. My hope is to rebuild logicism – in a way that preserves an apparently archaic idea. I will represent mathematics as a nonempirical discipline founded on basic truths of reason. This rationalism will require much qualification. It simply cannot be true as it would have been meant a century ago, and my updated version is speculative. But so is any philosophy of mathematics. I want to make logicism a leading epistemological option once more.

I also intend to identify logicism in my sense as a historical entity. The issue here is not whether arithmetical truths were truths of reason for Frege. Of course they were. But there is much disagreement over just how to interpret this idea; over its place in Frege's thought; and over Frege's relations to other foundational thinkers of his time. I want to stress certain aspects of a logicist position that Frege substantially shares with Dedekind and that sets both thinkers apart from later ones. I will call it (with apologies to Leibniz and Bolzano) first-generation logicism.

Since first-generation logicism can, as I will argue, resist standard objections while adapting to recent insights, the movement away from it was guided more by fashion than by logic. Insofar as we are products of that movement, my reconstruction will reflect on us as well as the early logicists. It should make us quite diffident about our rationality – an attitude anyway recommended for rationalists.

I

I will deviate from my position of 1987, where I restricted (deductive) logic to first-order logic (FOL). If logicism is true, logic must include a substantial set theory. But my intent is not so much to withdraw my earlier views as to explore an ultimately compatible alternative.

According to my 1987 essay (pp. 5-14), a logic defines the

patterns of ideal justification. Any logical implication of a set of premises can be derived from it in finitely many steps having certain epistemic properties. Roughly, each step is obvious and irreducible, and immediate consequence via such steps is decidable. I argued that this conception of logic was motivated by a rationalist vision of systematic, comprehensive theory; of a fabric of statements linked by justifications that beings with unlimited finite computational power could survey. The relevance of this vision for such limited beings as ourselves was, in turn, established by a social conception of knowledge (pp. 27–31). Human communities may be regarded as meaningful approximations to rationalist inquirers. Thus a communal, not individualistic, viewpoint suggests the identification of logic with FOL.

Although much in that paper needs elaboration, I will make just one comment. Rationalism is supposedly the party of epistemic reaction – of foundationalism. Thus, a rationalist conception of logic may appear outdated almost by definition. But rationalism in my usage involves no Cartesian foundations (any more than empiricism necessarily disavows them). A rationalist must hold pure reason, in some sense, to be a main source of knowledge. She must also accept an ideal of systematic theory; and with systematicity comes an interest in identifying foundations of our knowledge at any moment. But the foundations need not be certain, unrevisable, or complete. My view, then, was that FOL would be used to establish justificatory connections among the beliefs of an ideal reasoner. Logical derivations would not have to trace beliefs back to a fixed or self-evident basis. Nor did the interest of my epistemic characterization of logic presuppose such a basis. I advocated a brand of rationalism not excluding the main epistemological insights of this century (cf. Foley 1987).[1]

One question is whether my (1987) view is too narrow. My arguments for a first-order view of logical *consequence* left room for a more substantial notion of logical *truth*. Of course, logical truth is usually defined in terms of consequence, the truths being the consequences of null assumptions. But one might also hold that, while FOL fixes the language and entailments of mathematics, certain statements are part of logic without being first order valid. They might, for example, be fundamental to thought in a sense conferring logical status. One's broader formulation would then identify logic with the first-order closure of the relevant truths of reason – a definition still harmonious with my general orientation (rationalism, the use of epistemic criteria for logical status) of 1987.

Both the narrow and the broad view may be legitimate. In the good sense of "logic" I set out in 1987, logicism is false. But another conception, more congenial to logicism, might be defended. It would keep the first-order notion of consequence but propose a sense in which axioms of set theory (say) are logical without being valid.[2] This is the idea we will pursue.

Mine will be a less than thoroughly revisionary logicism. First, I accept what I take to be two major insights of contemporary mathematical epistemology: much of mathematics is not certain in any strong sense, and hypothetico-deductive reasoning, or (as I will say) explanatory inference, plays an important role. These insights have transformed the old picture of mathematics as advancing just by Cartesian intuition and deduction, and I would not turn back the clock. Second, my view is already widely held – up to a point. I believe, in essence, that mathematics proceeds by intuition, deduction, and explanatory inference. What we call theorems are, as ever, deduced from axioms. But while intuitive evidence is sufficient for some axioms, others rest on a combination of intuition and explanatory, systemizing power. And intuition itself is fallible. Now this picture is familiar: for example, from Russell and Whitehead (1910: x), Gödel (1947/64), and Putnam (1975). It also appears at the conclusion of Maddy's recent, authoritative survey of set theory (1988). Quine, too, endorses it – with the understanding that intuition simply drops out. We might well call it the standard model of mathematical justification.

But the agreement hides differences over both the nature of "intuition" and the data for the explanatory inferences. On the latter point, Quine refers all explanations ultimately to sensory evidence, while other writers allow a partially or totally nonsensory base. As for intuition, it may serve as a label for diverse noninferential means of cognition. The recent trend is toward grounding basic mathematical beliefs in perception. For the general trend in epistemology has been empiricist, and this is one way to guarantee mathematical empiricism (e.g. Kitcher 1984; Maddy 1980). Since it is agreed that deduction and explanatory inference do not lead outside the class of empirical beliefs, a perceptual reading of "intuition" would make all mathematical justification empirical. (In contrast, Quine maintains empiricism without needing any account of intuition.) Logicism, however, must reject this line. Explanatory inference is permitted, but the ultimate as well as the immediate explananda must be nonperceptual. And the relevant intuitions must be purely conceptual in a reasonable sense.

Thus, my version of the standard model is apriorist. Mathematics proceeds from *a priori* beliefs via deduction and explanatory inference – processes that preserve apriority. The basic beliefs, although at least sometimes fallible, are analytic. A modified traditional view of mathematics and proof will thus emerge. Mathematics is essentially a structure of proofs, which are deductions from axioms. I am traditional, and logicist, in insulating mathematics from observation and experiment. (To be accurate, this holds for computationally idealized beings.) The contemporary twist is that the axioms may be hypothetical as well as intuitive, and uncertain either way.

This essay will, admittedly, touch very lightly on actual mathematical practice. I will concentrate on the preliminary task of clarifying the thesis and blocking immediate objections. I will also pursue the historical questions raised above. The standard model of mathematical justification has been around at least since Russell. What distinguishes recent versions is, in general, the attempt to fit it into an empiricist epistemology. But it is just there that these models become forced and implausible. I want to appropriate the standard model for the logicism of Frege and Dedekind: a superior version of this model evolves naturally out of their views. Crucial here will be the view of intuition as being analytic, not perceptual. Empirically inclined philosophers have, in essence, tried to exploit the origin of mathematical belief in our perception and manipulation of objects. Yet that is to chase a mirage, as two great rationalists, Frege and Chomsky, have stressed in quite different ways. Virtually any belief has perceptual origins in this sense, but that has nothing to do with the epistemologist's questions of justification. I suspect that the empiricist strategies here have been driven by the conviction that one empiricist story or another must be true. Undermining that idea may cancel their appeal.[3]

My project here would be impossible unless some useful path through the myriad ideas relevant to logicism were already given. Fortunately there is one. Philip Kitcher has addressed the same questions at length, while drawing radically different conclusions (especially 1984, 1986, 1988). It will therefore be natural for me to develop my position in dialectical opposition to his, thus acknowledging, of course, the value of his work.

I will conclude the preliminaries by returning to a question I raised in my 1987 essay: given the diversity of extant characterizations and conceptions of "logic," how nonarbitrary can any particular choice be? This is central to the problematic of logicism, since sufficient ambiguity in our concept of logic vitiates any *argument* over whether mathematics is logic. In response, I advanced one conception of logic as being plausible and deeply interesting, while arguing that some main competitors were just unreasonable. Another (compatible) reaction is as follows. It is now unclear whether fixing boundaries of logic matters – whether the issue is more than verbal. To gain illumination, one naturally asks what the identification of mathematics with logic meant to Frege and Dedekind. Of course, nothing forces us to share their viewpoint. But lacking a definite reason to care about how logic's boundaries are drawn, we should consider the first and historically most significant answer. If anything can constrain our use of "logic," it might be the programme within which the modern notion of logic was created. And if we want to know whether the boundaries are worth arguing over, the original reasons for thinking so are the place to start.

II

The leading idea of first-generation logicism was that arithmetic (henceforth including analysis and algebra, unless the context shows otherwise) is analytic *a priori*. This claim was made in partial opposition to Kant, within his framework. Kant had distinguished the *a priori* and the *a posteriori*, the analytic and the synthetic; the logicists fundamentally accepted his epistemology but held that he had misclassified arithmetic. They believed that the arithmetical laws could be derived from logic and definitions alone, thus passing a Kantian test for analyticity.[4]

More accurately, the logicists did not assert quite what Kant denied. Kant's notion of logic was narrow, excluding relations and multiple generality. The logical revolution in turn transformed the notion of analyticity. Yet the logicists' Kantianism was genuine. The content of their analyticity claim was, on the negative side, that arithmetic requires no grounding in spatial or temporal intuition. On the positive side, foundations for arithmetic could be found in basic, universal principles of reason – laws of thought applying to any rational being at all, whatever her perceptual makeup. So the logicists upheld Kant's basic notion of logic and his diversion between logical and intuition. Logicism begins with a distinction between experiential and *a priori* sources of knowledge; within the second category, it distinguishes intuitive from purely rational sources. It then classifies arithmetic as nonexperiential, purely rational knowledge.

Here I am in effect identifying Dedekind's views with Frege's. The difficulty is that Dedekind is philosophically and logically inexplicit; locating Frege's viewpoint is much easier. Moreover, their ideas could hardly agree in every detail. But since we will shortly focus on possible differences, we may proceed with a provisional sketch of their common position.

One point of clear agreement for them concerns the purely mathematical motives for logicism. A derivation from logic would yield proofs of previously unproven arithmetical axioms, thereby reducing our primitive assumptions. For both men, this would be mathematical progress (Frege 1884: 1–2 and *passim*; 1893: e.g. VI; Dedekind 1888: preface).

So logicism is an epistemological thesis first: a claim that certain proofs exist and that their premises have a special epistemic character. But Frege and Dedekind include a distinctive ontological component closely tied to the epistemology. They are Kantian constructivists who diverge from Kant (and the intuitionists) by holding that numbers are constructed by pure reason.[5] This is meant literally: the human intellect makes, or creates, the numbers. Indeed, it does so as a special case of its ability to create sets. Numbers and sets are free creations of the human mind. In the Kantian tradition, such a view is tied to the epistemology. Purely rational knowledge of the numbers is made possible by the purely rational source of the objects, just as for Kant *a priori* knowledge of geometry is made possible by the transcendentally ideal character of space.

This constructivism at once obstructs and eases the defense of logicism. On the one hand, it adds a false component. I assume that if mathematical objects are real, they are no more constructed or mind-dependent than trees or planets. And if the logicists' epistemological thesis depends on a purely rational construction of the numbers, then rejecting the constructivism means rejecting their epistemology. On the other hand, one of the main objections to logicism now seems irrelevant. The ontology of mathematics is often cited against logicism: how could the *existence* of certain objects (worse yet, infinite structures) be a matter of *logic*? But evidently, nothing in the first-generation conception bars logic from introducing particular objects.

The reply will, of course, be that it is just the possibility of rational construction that makes the logical ontology admissible. Give up the constructivism, and you can no longer explain how anything could exist as a matter of logic: how reason alone could show us particular objects.

Only a modified logicism survives this objection. The first-generation attempt to ground the epistemology in a Kantian metaphysics was wrong. That grounding, however, was useless from the start. The idea that our constructive activity ensures epistemic access to the constructions is a recognized (although tempting) error. There is no quasi-Cartesian guarantee of knowledge about the products of our thinking, whatever the exact nature of the production might be. The construction of the numbers by the human intellect would imply nothing about the possibility of arithmetical knowledge. Insofar, then, as the logicists' epistemological thesis is still primary, another defense would be needed anyway. One should recognize the irrelevance of the constructivism and seek an alternative case for the analyticity of arithmetic. In doing so, one need not immediately give up the vision of a purely rational – in some sense – route to an ontology.

Let this suffice by way of introduction to logicism. I will elaborate by taking up the argument with Kitcher. We disagree over the interpretation and assessment of early logicism and, fundamentally, in our views of what has happened since. For Kitcher, Frege's and Dedekind's views are superficially alike but deeply opposed; the latter alone is said to have anticipated a correct naturalism and anti-apriorism (1986: *passim*). Thus, Kitcher sees a subsequent history of misguided steps taken under Frege's spell. In my view, however, Frege and Dedekind shared an important vision that has since been ignored and misunderstood.

III

Central for Kitcher is a critique of apriority. While agreeing that arithmetic was *a priori* for Frege and Dedekind, he finds this term ambiguous enough to require major clarifications (1986: 313–18).[6]

(1) Apriority as possession of an *a priori* warrant (APW apriority). The idea here is that a belief is *a priori* if it was produced by a process P that confers warrant independently of experience. Assume some way of segregating belief-producing processes into the experiential, or partly experiential, and the rest. *A priori* warranting processes, then, are nonexperiential processes that confer warrant even in the complete absence of experiential support. Now Kitcher interprets this idea in a strong way: for P to confer an *a priori* warrant, any belief produced by P must be warranted no matter what experiences might follow. *A priori* warrants can never be empirically overridden (p. 313).

For example, a mathematical proof provides an *a priori* warrant only if it is short enough to be grasped all at once, without loss of demonstrative force.[7] For if following it requires notes, diagrams and the like, then our conclusion rests in part on the experiences of inspecting such aids. It would be unjustified in a world in which these experiences were lacking or one abounding in illusory notes and pictures. Further, the proof cannot provide an *a priori* warrant if social pressures could cancel its force. Suppose that the mathematical community were publicly to reject many of the proofs you rely on. (Perhaps they come to have doubts about indirect or inductive proofs.) Then, Kitcher argues, it might well be irrational of you to maintain beliefs resting on those proofs. For given the advanced development of mathematics, individuals should defer to the considered judgment of an expert community. But this, according to Kitcher, leaves room for few *a priori* warrants. Almost any justification can be undermined by the experience of discovering that the experts reject it. Thus, APW apriority is out of the question for the great bulk of mathematics.

(2) *A priori* beliefs as propositional preconditions of thought (PPT apriority). Proposition P is PPT *a priori* "if and only if necessarily, for any rational being S, if S thinks (forms beliefs), then S must believe that P" (p. 314). A proposition might be APW *a priori* without being PPT *a priori* (pp. 315–16). For example, a proposition following by logic from certain beliefs rational beings necessarily hold might arguably have an *a priori* warrant, yet actually believing it might not be a precondition of thought. Indeed, all sorts of beliefs might be APW *a priori* without having any claim to PPT status at all. Even if the axioms of set theory were knowable by infallible, purely rational intuition, one would

not claim that only set theorists can think.

Kitcher, however, is particularly concerned to argue that PPT *a priori* beliefs need not be APW (pp. 314–15). He claims that there is no guarantee that what is PPT *a priori* should be known at all. Further, if such a proposition is known, our grounds might, necessarily, be open to experiential challenge. "It is eminently possible that these [PPT apriorities] should be identified by reflection on and analysis of our thought and discourse ... and that an unkind experience should override our right to suppose that our analysis had led us to correct conclusions" (p. 315). These points, which are central for Kitcher, will receive comment below.

(3) *A priori* propositions as mind-imposed (MI). "To say that a proposition is *a priori* may be to claim that it is true in any world of which we can have experience, and that its being true in any such world is the result of the structure which the mind imposes on experience" (p. 317). This third notion of apriority is obviously distinct from the others. Most important for Kitcher is the fact that an MI truth may not be known to us, let alone *a priori* warranted. We may have no idea what our mind imposes.

Recall that for Kitcher, no advanced mathematics to speak of is APW. Crucial to his reading of the historical situation, then, is the claim that while apriority means APW apriority to Frege, Dedekind's position differs in this and several other important respects. Thus, Kitcher can represent Frege as an epistemological reactionary, Dedekind as the wave of the future. In outline, his picture is this:

(A) Whereas mathematics is APW *a priori* for Frege, Dedekindian apriority consists in PPT status and MI (pp. 315–18). For Dedekind, mathematical knowledge is gained by examining existing mathematics and trying to find systemizing principles. These are justified by their utility for deriving accepted theorems (under such constraints as consistency and plausibility, of course). This is a hypothetico-deductive procedure with an essential empirical component, since it is an empirical matter what mathematics is accepted at any moment. Indeed, Dedekind would be prepared to allow that mathematics is "thoroughly empirical" (p. 318).

(B) These points shed light on Frege's attitude toward set theory, including his critique of Dedekind's set-theoretical foundations, and on the impact of Russell's paradox (pp. 324–7). The reason Frege cannot accord set theory logical status is that logical laws express universal prescriptions of rational thought – yet positing a universe of sets cannot be universally prescribed. Fregean logicism is therefore not confirmed by a derivation of arithmetic from set theory. And it is in trouble once Law V is refuted. For by what other principle could one try to introduce objects, if the introducing principle must be a universal element of correct thought?

Dedekind's task, however, is to systemize arithmetic using more general principles. These need not be APW. Their apriority is, rather, to be understood as follows (pp. 331–5). Certain mental operations – "collecting and [functional] correlating" – are necessary for any thought whatsoever. To employ these operations is to accept certain "logical" principles that let us create the numbers. In fact, these are principles of set theory, which is therefore PPT *a priori*. And the arithmetic derivable from set theory is itself *a priori* in this sense. The description of any intelligible world requires the resources of arithmetic; a thinker there will need to use the operations of collection and correlation to construct the numbers. Thus, thought requires arithmetic. Yet nothing about APW status follows.

(C) The apparent centrality and inescapability of arithmetic in our thinking is emphasized by Dedekind, whereas Frege only notes it in passing (pp. 335–6). The attempt to show that arithmetic is APW cannot illuminate this feature of arithmetic, but once we give that up, we may find the problem a fruitful one for cognitive science (pp. 338–9). Dedekind encourages us to explore the idea that arithmetic falls out of fundamental psychological operations, and that it is useful no matter what the world is like.

(D) Dedekind's constructivism, which might seem tied to an outlandish metaphysics, is naturally transformed into Kitcher's own position (pp. 328–31): that mathematics describes the activities of an idealized constructor. The transformation is natural in part because it seems the nearest way out of an obvious difficulty. Constructivism risks leaving us with purely subjective objects – why should the numbers I construct be the same as yours, Kitcher's, or Newton's? Kitcher's solution is to provide the same arithmetic for all by taking us each to refer not to our own constructions but to those of an idealized being (1984: ch. 6).

Kitcher's view of Frege and Dedekind, however, is quite untenable. The analysis fails at many points, notably in the account a apriority. I will mention three of Kitcher's historical distortions and then examine questions of apriority at length.

IV

(1) Frege's criticism of set theory is, *pace* Kitcher, unrelated to any project of providing APW foundations. As he makes explicit in Frege (1895), he opposes extensional foundations because he favors a concept-theoretic starting point; concepts are logically or conceptually prior to extensions. Note that this involves no rejection of set theory or denial of its logical status. The claim is rather that the concept-theoretic viewpoint is more basic. So deriving arithmetic from set theory is satisfactory, assuming a prior foundation for the latter.

Further, Kitcher's comparison of Frege and Dedekind on set-theoretic foundations is hard to follow. He argues that set theory (in something like Zermelo's style) cannot be PPT *a priori* (p. 327) and that this counts against Frege. Frege, Kitcher thinks, wants the foundation for arithmetic to be both APW and PPT *a priori*, thus "conflating" the two notions (pp. 326-7). But if the problem lies in the attempt to establish PPT status, and if apriority is supposed to be PPT apriority for Dedekind, why should this not be (at least) equally Dedekind's problem?[8]

Indeed, Kitcher's contrast is unreal. The set theory can, for Dedekind, be logical only if it is generated (in some sense) by fundamental operations of thought – collecting and correlating. But what is "collecting" if not abstraction? As Frege observes, it differs from Law V just in that Dedekind never explicitly states it (1893: 1-2; 1903: 253). So both men posit a fundamental operation that can give us a mathematical domain. The operation has the same status and workings for each. Neither man conflates anything, but Russell's paradox defeats them equally.

(2) Kitcher finds the essence of Dedekind's position (in particular, of his apriorism) best expressed by Frege, in the *Grundlagen* passage on the inconceivability of a nonarithmetical world (1884: 20-1). (He thinks that Frege achieved the successful formulation – "a lucid exposition of the thesis that Dedekind labored to explain" (p. 336) – even though what was Dedekind's central idea held only passing interest for him.) But Frege's reasoning there is confused. Hence, Kitcher's project of basing an alternative, essentially non-Fregean logicism on the idea of this passage is undermined.

Frege contrasts arithmetic with geometry, remarking that while we can consistently imagine worlds in which Euclidean (or any

LOGICISM

other) geometry fails, we can conceive no world in which the laws of arithmetic fail – "complete confusion ensues" (1884: 21). From this he infers that the laws of number should be intimately connected to the laws of thought. But the contrast is erroneous. For all Frege shows, the difference between arithmetic and geometry lies in the fact that just one of these is a theory of a particular mathematical structure. A variety of spaces is conceivable, because the notion of space (line, plane, etc.) is not tied to any one geometry. Arithmetic, on the other hand, is the theory of positions in a simply infinite sequence – the number series. Thus, Frege has set up a false comparison. What corresponds to geometry is not arithmetic in the ordinary sense, but rather the theory of ordered structures. Just as we conceive of non-Euclidean geometries, so we can conceive of ordered structures that violate arithmetical laws. Any structure not isomorphic to omega will do. In fact, we speak of deviant or variant arithmetics just as we speak of variant geometries – think of the arithmetics of complex numbers, infinite ordinals and cardinals, modular systems, and so forth. Once the cases of arithmetic and geometry are properly aligned, no obvious difference remains. A variety of geometries and a variety of generalized arithmetics (theories of ordered structures, or of certain species of these) are conceivable. When we focus, instead, on a particular structure, either arithmetical or geometrical, we cannot conceive it without its defining features. There can be no whole numbers between two and three and no intersecting parallel lines in a Euclidean plane.

Does this just show that geometry might, contrary to what Frege thought, also be logic? Not unless one wants to construe "logic" broadly. Married bachelors, for example, seem just as inconceivable as numbers between two and three or intersecting Euclidean parallels. Is it a matter of logic that no bachelors can be married? That has been maintained, but if that is what "logic" means, and if arithmetic is logical in that sense, then Frege's passage no longer illuminates the nature of arithmetic. It fails to explain a special status for arithmetic (numbers are not shown to be relevantly different from bachelors or triangles), and it fails to connect arithmetic to laws of thought in any clearly significant sense.

Besides, Kitcher's main reason for taking Frege (1884: 20–1) to be peripheral seems to be its lack of fit with his general view of Frege. We will develop a different picture.

(3) It is misleading, in this context, to emphasize *Dedekind*'s

constructivism. As already noted, Frege was no less constructivist. This is not entirely a small point, for if we aim to understand first-generation logicism, and to compare the implications and prospects of its two main versions, then recognizing this fundamental agreement on metaphysics is important. Moreover, Kitcher's remarks on constructivism are anachronistic, showing little feeling for what is distinctive in the early logicist position.[9]

He recommends "[giving] up the misleading and problematic idea that our mental operations bring into being certain objects as their products. Instead ... mathematical statements [should] be construed as describing the properties of the operations themselves" (p. 329). But, granted that the constructivism is wrong, simply dropping this key element and moving to Kitcher's own theory is hardly in the logicist spirit. If we can retain anything of logicism without holding that numbers are created, it would be the idea that there are arithmetical objects and that these can in some sense be recognized purely by reason. For within a Kantian context, the constructivist view of mathematics was taken for granted. The question was whether the constructions occurred in intuition or elsewhere. The real opposition came from thinkers offering radical reinterpretations of mathematics, for example, as involving the manipulation of meaningless symbols or as describing mental processes. Logicism held, with Kant and against the formalists and empiricists, that arithmetic dealt with mathematical objects; against Kant, that arithmetical knowledge was purely rational. If the constructivism is omitted, what remains is some form of platonism that posits a nonempirical, nonintuitive access to mathematical reality. I sympathize, but the view is anathema to Kitcher.

As for the problem of subjectivism, Kitcher's way out would be alien to Frege and Dedekind.[10] The Kantian solution is to posit a species of content ("objective content") determined by the shared structural features of subjectively different constructions. I explained Frege's development of this line (in 1983a), and Michael Friedman (especially 1987) has recently shown the importance of similar ideas in Carnap (Frege's student) and other figures of our century. Although Dedekind's philosophy on this point (as elsewhere) is not explicit, he, too, would presumably follow the Kantian line.

The points of this section show the pattern of Kitcher's interpretation: he tends to exaggerate the differences between Frege and Dedekind while underestimating both the latter's problems and his distance from Kitcher's own position. This pattern, as we will now see, is most pronounced in Kitcher's discussion of the key issue of apriority.

V

Let us begin with the relations among Kitcher's various species of apriority. Clearly, an *a priori* warranted belief (in any reasonable sense) need not be a PPT. Even a self-evident truth might be one we could perfectly well live (and think) without. But it is questionable whether a PPT could lack an *a priori* warrant, even in Kitcher's strong sense. If Q is a PPT, then by definition, any thinker S believes Q. Thus also, belief in Q will persist no matter what experiences, representing social challenges or anything else, S undergoes. What Kitcher must therefore claim is that holding onto Q (even though one cannot help it) will be irrational in the face of a suitable social challenge: that the challenge will remove the warrant, although not the belief. But this stance is problematic.

It is not entirely clear how a challenge undermining the warrant should arise – any experts challenging Q must also believe Q. But perhaps some such group wishes to deceive S about Q, or perhaps it suffices for Kitcher if S merely believes that an expert community rejects Q. The problem is that S will find herself unable to renounce Q. Whether her continued acceptance of Q is warranted then becomes a complex question, one different epistemologies will approach differently.

From a traditional viewpoint, Q would seem to be a kind of self-evident truth. Retaining Q in the face of apparent expert rejection is not, then, clearly irrational, particulary given the least sophistication about experts. It hardly takes a Cartesian to suppose that an error by the authorities, a deception, or just a failure to understand them may be at least as likely as the falsity of what appears self-evident. Appearances of self-evidence may, of course, change with time. But the question is whether to reject Q when it does look self-evident, instead of supposing that the authorities are wrong or that they only seem to be rejecting Q. There is reason to stick with self-evidence.

A more modern epistemological turn will not help Kitcher's case. Pragmatists, for example, should regard PPTs as paradigms of justification. If giving up Q makes thought impossible, then the pragmatic move is to revise one's view of the experts in a way that does the least damage overall. The case is even clearer within Kitcher's own, reliabilist epistemology – assuming that a PPT must be true. For if PPT status guarantees truth, then holding on to PPTs in the face of expert challenges is a completely reliable method. The experts will always be wrong, the PPT always justified.

Although I have only sketched a few options for treating Q, and only in the case of social challenges, these remarks may generalize. If believing Q really is a precondition of thought, it is hard to see how any experience could make Q unreasonable. Thus, PPTs should be *a priori* warranted even as Kitcher understands this notion. (Would not his mathematical PPT beliefs be counterexamples to his own theory?) Of course any weaker understanding would make it still harder for PPTs to lack APW status.

Kitcher's views on MI apriority and the utility of arithmetic are equally problematic. He rightly notes that a truth imposed by the mind (assuming that even makes sense) need not be knowable at all, let alone APW. But then how is he able to construe MI as an *epistemological* category, as a kind of knowability? Kitcher's view is that, in any experienced world at all, arithmetical concepts and principles will be useful. But this leads to difficulties on Kitcher's own terms.

On the one hand, mind-imposition in this sense seems to provide an *a priori* warrant. No matter what experiences we have, including experiences of social challenges, it must be useful to do arithmetic. Thus, doing (and believing) arithmetic should be warranted no matter what our experiences, which is what APW apriority amounts to. Here Kitcher might reply that arithmetic has only been shown (or hypothesized) to be *useful* in any experienced world. Useful tools can be given up – capitulating to the experts might be even more useful. But this reply weakens the mind-imposition claim, which was presumably not just that arithmetic would always have some degree of utility. In short, if Kitcher is *serious* about MI, APW status for arithmetic is hard to deny.

On the other hand, MI as Kitcher explains it is dubiously consistent with the PPT status Kitcher favors for arithmetic. To say that it is always useful to posit arithmetic is to imagine rational beings deciding to accept arithmetic for its utility. Since this presupposes an ability to make the decision without already believing arithmetic, arithmetic cannot be a PPT – even if thinking beings in all worlds would have compelling reason to accept it. (We find just this picture in Dedekind. If numbers are free creations of the human mind, then there can be no strong necessity to think arithmetically. Arithmetic cannot be a PPT for Dedekind.)

Now such difficulties arise from taking Kitcher's explications of apriority at face value. But those explications are misleading with respect to what seems really to be his position – one he also attributes to Dedekind.

Kitcher proposes (i) that mathematics typically proceeds by trying to identify general laws that underlie existing mathematics (e.g. that systemize our present theories (pp. 314–18); (ii) that the laws uncovered in this process are (or can be) *tacitly* known by us all along (ibid.); (iii) that such laws are, in fact, tacitly known by all rational beings (pp. 334–6, 338–9); (iv) that they will be applicable in any thinkable world (ibid.). One conclusion Kitcher draws is that mathematics is empirical. Mathematical inquiry begins with the survey of existing doctrine – an empirical procedure. Moreover, the question what we tacitly believe is empirical. Yet Kitcher also intends to uphold a form of apriorism. For if (iii) obtains, then the laws we discover empirically are PPTs. And (iv) expresses a form of mind-imposition.

This position is (at least) doubly confused.

First, it is not really a way of maintaining PPT apriority without APW apriority. The apparent separation of PPT from APW status turns on an equivocation. When Kitcher claims that PPTs need not be APW, he means that tacitly believing Q could be a precondition of thought, even though one might have no reason explicitly to believe Q (p. 315). But still, the tacit belief might be warranted as a belief and immune to empirical refutation, hence *a priori*. Indeed, we have observed that PPTs would almost certainly be APW. So Kitcher has not really defended a lack of *a priori* warrants for PPTs at all. He has rather maintained that some PPTs need not be explicit. While this may have some interest, it is quite orthogonal to the question whether Frege used the wrong notion of apriority.[11]

A second mistake underlies the claim that mathematics is empirical, since only empirical inquiry can uncover the presuppositions of thought. Even granting that finding such presuppositions is an empirical task (we have no idea how to do it now), this is a levels confusion. What follows is that empirical inquiry is needed to establish the apriority of mathematics. Perhaps only cognitive science can show a certain mathematical proposition Q to be among the preconditions of thought. Thus, we establish empirically *that Q is a PPT*, but of course that shows no empirical status for Q. (Similarly for MI instead of PPT.) It is the epistemic proposition, not the mathematical one, that is empirical here; one can quite well hold that the mathematical Q is *a priori*, although its apriority is an empirical discovery. To suppose that mathematics is in the business of identifying presuppositions of thought *as such* is a muddle Frege would have noticed at once.

We see again the effects of Kitcher's attempt to force Dedekind out of his natural alignment with Frege toward Kitcher himself. Since Kitcher believes that mathematics is empirical, he must find an apriorism that is consistent with empiricism and that he can attribute to Dedekind but not to Frege. At the same time, he must attribute to the latter an apriorism the former does not share. Yet he fails to show apriority for the mathematical propositions, and the effort to find a difference between Frege and Dedekind involves confusions over the relation between APW and PPT apriority.

But subtracting these mistakes still leaves Kitcher with a substantial position. He presents a serious case against mathematical apriorism (especially 1984) and has developed a "mathematical naturalism" in opposition to *any* traditional epistemology of mathematics (1986, 1988). Moreover, the claim that Dedekind but not Frege anticipated mathematical naturalism may mark the principal difference for Kitcher. So we still have to deal both with a forceful challenge to logicism and with the possibility of a largely illusory unity within early logicism after all.

VI

Logicism without an apriority thesis is not logicism. Any logicist position must posit a nonexperiential source of warranted belief. Further, this source must generate, if not all mathematical knowledge, at least a substantial amount of it. Kitcher counters that mathematics is learned from experience and that the apriorist goes even farther wrong by rejecting the possibility of empirical undermining. But he overlooks reasonable options for the apriorist.

Apriorism begins with a distinction between experiential and nonexperiential beliefs. The division cannot be exclusive, since some beliefs must have mixed origins – as when an inferred belief has premises of both kinds. But the apriorist, having introduced this division, will assign certain beliefs to purely nonexperiential sources. Now she must further claim a suitable degree of *warrant* for certain nonexperiential beliefs. But her first move is already problematic. Although positing nonexperiential beliefs is a commonplace of our tradition, it involves fundamental difficulties.

One question concerns the idea that belief has sources at all in any traditional sense. Thought does, obviously, depend on sources of *information*. Cognitive science investigates the processing of olfactory, visual, proprioceptive, and other kinds of information. Information is transformed, stored, sent, retrieved, and so forth. But this does not easily translate into talk of sources of *belief*. When you look toward the mat, you are, in some *relatively* clear sense, obtaining visual information from the cat. You see the cat on the mat. But is the belief visual? Visual input helps bring it about, but there would be no belief without some complex integration of various kinds of information. The entire process is quite obscure. Generally similar remarks hold for the beliefs we are inclined to assign to other sensory sources or, say, to memory or reasoning. Lacking an advanced cognitive theory, we hardly know how much sense commonsensical talk of the sources of belief really makes. And it is unclear both how far cognitive science will illuminate such ordinary notions as belief and how well anything resembling the traditional division of the mind into different sources of belief will survive.[12]

Even if some form of that division survives, the idea of a nonexperiential source of beliefs may not. Perhaps all beliefs are reasonably labeled experiential, even if we further distinguish sources. Beyond that, there is a conceptual problem. Discussion of *non*experiential (or nonperceptual) belief requires a notion of *experience*, but what, in general, is *that*? Neither the enumeration of known modes of perception (vision, echolocation ...), nor some problematic claim to the effect that experience is just what generates qualia, is satisfactory. Current perceptual psychology may suggest a kind of answer: *very* roughly, that experience involves the classification and integration of incoming signals in a way that yields representations of their causal sources. But of course any such answer raises its own difficulties.

The moral is that the first problem with any apriorism is not whether it is true but whether its underlying model of belief is at all close to the mark. The apriorist (and not she alone) is committed to a highly general, vague hypothesis about the categories of an advanced cognitive science.

But to recognize this situation is not to dismiss every apriorist thesis. Lacking, for the time being, a clear improvement on the language of traditional epistemology, we may formulate a working apriorist conjecture. The claim would be that beliefs are generated by different sources roughly as epistemologists have thought, and that nonexperiential origins of the right kinds can confer warrant on mathematical beliefs. Since the point of this claim is the existence of certain cognitive mechanisms, let us call it *cognitive apriorism* (but often just "apriorism" when confusion is unlikely). We will find it useful in spite of its vagueness.

It is worth noting that talk of discrete belief sources is deeply entrenched. This is particularly clear when we consider, not internal, but external sources. We find ourselves able to tell what we have learned from this friend, from that one, from the evening news, *The Times*, and so on. If tracing beliefs to sources makes sense at all, it seems prima facie intelligible and reasonable to do so for sources inside the person – including, perhaps, sense modalities and various *a priori* faculties. Of course a future psychology may heavily revise our present models of the character and interactions of such sources. Still, even a revisable common sense is worth agreeing with.

Since cognitive apriority is intended to be close to mathematical apriority as traditionally conceived, it is a natural counterpart to Kitcher's notion of APW apriority. Both are intended to clarify the same idea. Yet they are quite distinct. Essential for Kitcher is the immunity of *a priori* beliefs to empirical undermining. Cognitive apriority entails nothing of the kind. The cognitive apriorist claims that warranted mathematical beliefs can arise *a priori*, that is, from nonexperiential sources. This leaves it open that beliefs with a very weak warrant, a moderate one, or at least one providing less than certainty can likewise so arise. In fact, cognitive apriorism as such does not entail that any certainties will arise at all. But then empirical undermining is in principle possible. This is fundamental: the properties of arising *a priori* and of having some kind of certainty (indubitability, unrevisability, or the like) are completely independent. Nothing in the nonexperiential character of the source guarantees any epistemic virtue.

Which is the right notion of apriority? Immunity to empirical undermining is undoubtedly a traditional element. Nor is this due just to what may be an implicit traditional assumption (which I am inclined to posit) of cognitively idealized agents. Even they will entertain hypotheses with various degrees of confidence. Hence, *a priori* sources of warranted belief might, sometimes or always, also provide them with uncertain beliefs. (Computational ideality is a long way from omniscience.) But claiming empirical immunity was a mistake.

The traditional view is understandable, given the search for certainty. If there are good reasons to find the senses uncertain, then one will naturally seek *a priori* sources. Any discovery of *a priori* propositions that do seem immune to doubt will encourage this enterprise. One may well come to think both that *a priori* propositions in general are certain and that a substantial body of *a priori* knowledge is possible. But wishful thinking aside, it is hard to see why nonexperiential origins confer certainty. And Kitcher's rejection of the alternative seems to involve a mistake of his own.

For Kitcher, empirical doubts could not threaten a genuinely *a priori* belief. (If every possible doubt can be communicated through an empirical source, this amounts to claiming that *a priori* beliefs must be certain.) The argument is simply that, if empirical information might bear on our acceptance of a belief, then there is no good sense in which it could be called independent of experience (1988: 321). But while this may be incontestable as a definition, it is irrelevant to an apriorism that deals with the sources of belief. Cognitive apriorism posits nonexperiential warranting processes. From this viewpoint, denying that experience could bear on *a priori* beliefs would be like denying that you could learn anything from your brother. (For no matter what your brother tells you, your sister might provide information that casts doubt on it.) This is contrary to perfectly reasonable ordinary usage. The point is that a belief can have a given source while still being revisable in the light of information from elsewhere. Certain *a priori* beliefs might also be unrevisable, but that would be a further question.

Kitcher might be little moved. For even if *a priori* knowledge in a good sense were possible, he would claim that any extensive *mathematical* knowledge needs empirical support. Not only are the proofs too long and numerous to keep in our heads, but we need other mathematicians to systematize and survey a body of information no one can grasp in detail. I share these views and grant that each of us can at best know a bit of mathematics *a priori*.

But here a theme of my 1987 paper is useful.[13] Recall the claim that recognizing FOL as logic depends on an interest in the justifications available to computationally idealized versions of ourselves; and that considering such beings seems relevant because epistemic communities can meaningfully approximate them. We let the primary knowers in our epistemology be communities, with their temporal resources, their libraries, and their computers. These knowers can, as needed, afford to connect their mathematical beliefs by derivations in FOL. Thus FOL provides the (deductive) logic for their practice of justification. In this way a concern with idealized knowers underlies our inquiry into the boundaries of logic. It is reasonable to ask not just what we can know *a priori*, but what they can.

Idealized knowers have unlimited (finite) capacities for storage and attention. They also avoid computational errors. Hence they need no pencils (or supercomputers) and paper, nor need they keep textbooks and other surveys of their field on hand. Thus, a great deal of mathematics might be *a priori* for them. If they can generate mathematical knowledge nonexperientially, they can keep extending their mathematical knowledge just by reasoning, without empirical crutches.[14] Since they have no problem with long proofs, *their* mathematics may be *strictly* (cognitively) *a priori*.

Why are we interested in what might be *a priori* for unreal beings? The situation is somewhat analogous to the use of idealizations in physics or of a competence–performance distinction in linguistics. We are interested in an epistemic classification of mathematical theory. Given our computational limits, almost all of mathematics is empirical for each of us. But, as I have argued (1987: 30–1), our particular finite limits seem arbitrary (unlike, say, the difference between finite and infinite limits, which seems deep). They introduce a kind of noise that obscures a more revealing picture, perhaps one showing important differences between mathematics and other sciences. If the idealization lets such a picture emerge, its introduction is justified.

And again, there is a way to connect the idealization to actual practice. One may extend the distinction between *a priori* and experiential belief to communities of nonideal inquirers to propose that much of mathematics is *a priori* for human communities although not for individuals.

Allow that individuals can gain warranted mathematical beliefs by nonexperiential means – by reflection, for the sake of a label. Then mathematics may work somewhat as follows: individuals come to believe some axioms by reflection. These are made public and collectively examined for reflective acceptability and coherence with existing mathematics. If they survive, they become part of publicly recognized mathematics, thus counting as communal beliefs. Theorems are deduced, leading in time to further reflections on the concepts arising in mathematical investigation. Here the mathematics is *a priori* in the sense of being justified strictly by reflection plus procedures that ideal agents would conduct without empirical aids. Since examining chains of reasoning and checking a result for fit with established doctrine are such procedures, a community's mathematics could be *a priori* in this sense – communally *a priori*. What communal apriority requires is that some mathematics be genuinely (cognitively) *a priori* for individuals and that the community's main contribution should be to overcome the individual's computational limits. (It may also play similar roles, such as providing a great variety of reflections, of which those that best withstand criticism survive.)

In short, mathematics may be cognitively *a priori* for ideal agents and, under a suitable extension of that notion, for communities. This is still a weak apriority thesis. If mathematics is the fallible product of *a priori* sources, then empirical confirmation or infirmation of this fallible doctrine is in principle possible. The empirical data might settle questions that *a priori* procedures leave open. So it would appear contingent whether any actual mathematics has been empirically tested. If it has been, then it is no longer *a priori* in any interesting sense. Some of it may have originated nonexperientially, but it has now become empirical. And Kitcher would claim that, if the absence of empirical testing is at best an accident, then the empiricist is winning.

A stronger apriorism would therefore somehow insulate the mathematics generated *a priori* from future empirical challenges. Although it is unclear how far we can go in this direction, there is, at least, a way to deflect the threats Kitcher stresses. A community's mathematics is, as I will now argue, immune to social challenges.

Kitcher's fundamental idea is a surrender of the autonomy that characterizes traditional knowers. Expert rejection should make one give up mathematics that is based not on hearsay but on clear proofs. Nor need a proof be questioned because of its length. The problem of social challenges is supposed to be additional to the problem of long proofs. Even someone who has grasped a proof in its entirety and in all its details should suspend belief if the experts reject the argument. Now this *may* hold for the beliefs of an individual confronted by a community of experts. But could a well-established communal belief be overthrown in the same way?

Suppose we meet the Galaxians, with their futuristic technology, dazzling intellects, and a flourishing mathematical community. And they deny the Gödel–Cohen consistency theorems. They are unable to demonstrate, to our satisfaction, errors in our proofs. That is, they cannot provide what *we* recognize as *mathematical* reasons to change our minds. This is the test case for Kitcher: the question is whether to drop our theorems on the word of authority alone, without independent reasons. I suggest not. Why should our community, with its own successful practice, give up major results it apparently understands in favor of conclusions for which it can see no mathematical support? The only argument is that the Galaxians must be right because they are so much smarter. But while this may be a good reason to use Galaxian machines, it does not settle the mathematical question. Maybe "set" or "proof" means something subtly different to them, in a way that they have failed to communicate when trying to explain their advanced mathematics to us. In that case their denial of our theorems also reflects a misunderstanding. They are wrong about our meanings, and theorems, even though they know far more mathematics.

We should not disrupt our science by rejecting apparently solid proofs of important theorems. The only possible concession would be to adopt some instrumentalist or fictionalist stance: continuing to accept our theorems while granting that somehow, for reasons we may one day see, they must be wrong.[15] But given the possibility of subtle failures of communication, we have a right to leave our mathematics alone. The chance of misinterpretation would always seem to be higher than the chance of a genuine mathematical error on our part (once we have checked our proofs). Indeed, the only way for the Galaxians to show that they have not misunderstood us is to explain just where and how our proofs (or axioms) are wrong. But then we will have mathematical reasons to change our minds. Deference will be unnecessary.

The arguments for mathematical empiricism of Kitcher (1984)

invoke long proofs and social challenges. Since the beliefs of a mathematical community resist social challenges, and since libraries, computers, and the like overcome the problem of long proofs, mathematics still appears to be communally *a priori.*

I must now emphasize that my apriority thesis for mathematics leaves much of Kitcher's aposteriority thesis in place. The two claims begin with different notions of apriority, since I substitute cognitive apriority for Kitcher's much stronger notion. And while Kitcher deals with individuals, I am speaking of communities. Moreover, communal apriority simply defines away one of Kitcher's *a posteriori* elements. Beliefs may be communally *a priori* even though they depend on written records. But my aim is not so much to refute Kitcher as to better understand the nature of mathematics. Without denying the ways in which mathematics may be empirical, I want to locate mathematics within a more illuminating scheme of epistemological classification. To that end, it is worth suggesting that mathematics is not empirical for computationally ideal finite agents, and that the ways in which it is empirical for human communities are correspondingly limited. This would, among other things, mark a clear difference between mathematics and natural science. No one thinks that computationally ideal agents could do physics nonempirically.

Further reflections on the possibly nonempirical character of mathematics will follow shortly. I continue by clearing mathematical apriorism of what may be Kitcher's most serious charge.

Kitcher opposes a view he calls mathematical naturalism to any traditional epistemology of mathematics and particularly to apriorism (1988). Naturalism holds that the development of mathematics at any point depends on the mathematics that already exists. What is given will set our problems and will constrain solutions. Further, new mathematics is, specifically, justified by explanatory inference. We look for axioms that organize and explain results that were previously not clearly connected or that lacked generality. Axioms that advance the field in this way without generating contradictions (or other particularly implausible results) tend to become accepted mathematics, and the process starts anew. Let us call the first naturalist tenet the historicity claim[16] and the second the claim of hypothetico-deductive procedure.[17]

Historicity is an obvious fact, and, although Kitcher's stress on explanatory inference may be a bit one-sided, he and others have established its importance.[18] But the supposed opposition between naturalism (in Kitcher's sense) and apriorism is illusory. Even a rigid apriorism, recognizing only indubitable intuition plus deduction, can admit historicity. First, no apriorist holds that we can simply decide any mathematical proposition whatsoever. Only a few answers are immediately evident at any time; the rest would need proof. And for us, what we can prove depends on what proofs we already have. We see only a few steps farther at any point. So the available, historically contingent collection of proofs determines what we can next apprehend. Second, there is no guarantee that we will intuit any proposition Q that is potentially self-evident to us. Maybe Q will be obvious *if* Q is properly considered. But what if we fail to consider Q? Perhaps we only ask the questions that seem pertinent in the light of our present beliefs, needs, and other elements of our historical situation. Third, apriorists can happily leave a place for conceptual development. One may *consistently* hold that Zermelo's axioms were self-evident for him (even if they actually were not) but that no one much earlier was in any position to intuit them. The work of Cantor, Dedekind, and Frege may have been necessary in order to provide a notion of set, including a working experience with it, from which to draw intuitions. For (at least) these three reasons, anyone can admit a strong historicist thesis. What we know *a priori* at any moment may depend heavily on the results, questions, and concepts generated by mathematical history to date.

The first component of Kitcher's naturalism therefore has no connection to the specific epistemology of mathematics. We can ask how new ideas arise from their predecessors in any evolving discipline at all, whatever its subject-matter, foundations, and methods.[19] In contrast, the claim of hypothetico-deductive procedure is an important challenge to traditional epistemologies of mathematics. But it is irrelevant to *our* apriority question. If the class of *a priori* beliefs is characterized not by certainty but by origins in nonempirical processes, then it may be expanded by explanatory inference. Traditional apriorists agree that deduction proper may lead from *a priori* beliefs to *a priori* beliefs; having given up the insistence on certainty, we can accept nondemonstrative modes of reasoning as likewise preserving apriority. Employing a different pattern of inference does not mean appealing to experience.

So far, our apriorist hypothesis is as follows. Mathematics is *a priori* in the sense that warranted mathematical beliefs can originate nonexperientially and that ideal cognizers can develop mathematics by nonexperiential means. This sense is compatible even with a complete absence of certainty from mathematics and with a dependence of *a priori* cognition on historical circumstance. Although individual humans have limited capacities for *a priori* mathematical knowledge, human communities can do better, practicing a mathematics that is empirical only in its use of aids to memory, communication, and computation. Among other things, communal mathematics resists social challenges.

Kitcher also mentions the possibility of what he calls theoretical challenges. This is just Quine's idea: that mathematics might be revised in order to simplify our total empirical science. But while this might be true, no good argument for it has ever been given. Quine's general argument for the empirical character of all belief assumes a discredited behaviorism (1960), and historical studies have not so far borne out his view of mathematics.[20]

But we recently omitted the answer to a simpler empiricist challenge. If some beliefs generated *a priori* are uncertain, might not empirical evidence raise or lower our confidence in them? Might not mathematics thus become empirical? Indeed, traditional apriorism recognizes empirical support. A mathematical proposition can be conjectured on empirical grounds, and the scientific application of mathematics seems to provide a systematic empirical basis. The standard apriorist reply would be that, although a mathematical proposition may receive empirical support, it is not part of *our mathematics* until it has been *proved* – deduced from unquestionable axioms. But now we are modifying this notion of proof. Although proof is deduction from axioms, the axioms need not be established by infallible intuition. So why not admit that even mathematical propositions originally warranted *a priori* (let alone any others) should confront experience?

The alternative is to hold to a modified traditional answer, roughly as follows:

1 Propositions must still be proved in order to enter accepted mathematics, and proof is still deduction from axioms.
2 Axioms in turn are accepted on the basis of some combination (which may vary with the case) of intuition and explanatory inference.
3 An axiom's explanatory power must suffice even if its (perhaps indirect) empirical support is disregarded.

In other words, what becomes part of mathematics must be acceptable on *a priori* grounds alone. To be accurate, of course, such grounds would be *a priori* only for an ideal being with no need for journals, blackboards, and computers.[21] Imagine an ideal reasoner, beginning with *a priori* intuitions and proceeding by explanatory inference, deduction, and further intuition, without relying on science. Let us say (tendentiously!) that she relies on *strictly mathematical* reasons. (Her mathematics could be called the *a priori closure* of the original intuitions.) If each of our theorems can be established strictly mathematically, then scientific confirmation is superfluous. Science may serve various heuristic functions, but we do not ultimately let it decide mathematical questions.[22]

Is this true? The question is both descriptive and normative. Clearly, the apriorist is describing a possible community. Mathematicians could proceed essentially without empirical tests; yet others, in other worlds, might not. Hence one question is how *we* proceed. The other is whether proceeding *a priori* would not show a snobbish, not to mention irrational, disdain for perfectly good data. This question has force both because we no longer take *a priori* origins to confer certainty and because *a priori* cognition might in any case have limits that we could empirically transcend.

The descriptive question is a prime target for naturalistic study, Kitcher-style. One would need to look in detail at how mathematical axioms come to be accepted; at the role or lack of one for empirical confirmation. These are open problems. At first glance, however, the apriorist has no obvious reason to fear the investigation. In fact, she may offer a quick plausibility argument. Maddy's survey (1988) makes it clear that set-theoretic axioms are now evaluated on mathematical grounds (in our sense), not with reference to empirical data. Since the rest of mathematics follows by logic from set theory, we have an argument for mathematical apriorism in general.

The reply would be that set theory rests on a body of analysis and geometry learned from experience (Maddy 1988: 761). It is just that those stretches of mathematics are now so familiar that the set theorist takes them for granted, perhaps missing the implicit empirical dependence of her rarefied reflections. But this is questionable. Insofar as an analysis of iterated collection, plus allied considerations, gives set theory convincing support, any empirical ladder we might once have needed can fall away.

Moreover – although this is a hard issue – even Kitcher's account of an earlier history, the development of analysis up to 1900, *seems* to be a story of mathematically driven advances (1984: ch. 10). Physics set many questions, but the mathematical hypotheses were established by their ability to solve problems within pure mathematics. Empirical support (from applications in mechanics) obtained but was dispensable. Moreover, the job was not finished until the desired results were rigorously derived from intuitively clear and plausible foundations – as the apriorist would have it. Thus, Kitcher's own story may be congenial to our brand of apriorism; and Maddy's account of the present scene certainly is.

Actual mathematical practice may, if it *is* apriorist, embody some wisdom. The general argument that it is unreasonable to pass up data, whatever its origin, meets a plausible reply of equal generality. Significantly expanding the data increases the risk of error along with the possibilities for discovery. But mathematical errors can be well hidden, one reason being precisely the distance of much mathematics (particularly now) from empirical applications. We can hardly expect a contradiction to appear each time a false theorem is accepted. Thus, mathematicians have reason to proceed conservatively, substantially accepting the discipline of axioms and proofs. The axioms may have hypothetico-deductive support, but such support is judged very conservatively and requires a sufficient supply of confirming instances that are beyond doubt.[23]

In this way, I follow Kitcher on mathematical apriorism after all. I conjecture that, under idealization, mathematics is immune to empirical tests. But it is not immune to *a priori* refutation. Further – and ironically – the reasons for shielding it from experience involve not arrogance about our *a priori* faculties, but the opposite. We can so easily fail to detect mathematical falsity that the risks of a freewheeling mix of empirical and *a priori* procedures outweigh the gains.

Whether this is really so will be a most intricate question. We need a careful examination of mathematical practice and of the rationality of its various elements. Kitcher should agree, since that is just what he thinks the philosophy of mathematics should be doing. But where he proposes this kind of study as an alternative to traditional concerns, and as a reaction to the death of apriorism, I regard it as a way of continuing the traditional argument. And the apriorist should not yet expect to lose.

I have been developing a form of apriorism as a step toward logicism. To proceed, we must investigate, first, whether our version is at all relevant to Frege and Dedekind's apriorism; second, just what nonexperiential cognition might be; and third, whether anything can be made of the logicists' analyticity claim. I will sketch preliminary answers.

VII

Frege may well be a Kitcherian apriorist. Logical laws should, for him, be self-evident, and proof should place a proposition beyond doubt (e.g. 1884: 2; 1903: 253). Moreover, Frege holds that support from a "mass of successful applications" is insufficient (1884: 1), thus rejecting a main point of Kitcher's. But the early pages of *Die Grundlagen* (1884) and (1893) (for example) show less interest in unassailable warrants than Kitcher indicates. Frege's aim is to find analytic *a priori* foundations. The issue is not so much certainty as the possibility of grounding arithmetic outside of intuition and sense experience (including psychology). And this is a matter not of degree of warrant but of the cognitive processes that may yield justified arithmetical belief.

It is true that Fregean foundational inquiry in general aims to rule out hidden inconsistencies (for example, 1884: IX). But this seems irrelevant to the investigation of number theory. Frege shows not the least concern about *its* consistency; in fact, that worry is dismissed. Frege observes that the basic laws of arithmetic are so amply confirmed by their countless applications that any request for proofs seems almost ridiculous (1884: 2). Seeking foundations in order to increase the warrants of (elementary) arithmetical beliefs thus appears to be a waste of time. Indeed, recall Frege's claim that, if we even try to deny these, such confusion ensues that thought becomes completely impossible. What greater warrant could anything have? Since the quest for proofs proceeds anyway, certainty is not its main goal.

Kitcher (1988: 320) emphasizes Frege's remark that without proper foundations our certainty is "merely empirical" (1884: IX). But we have just seen that insufficient certainty in number theory, at least, is not a worry. (And for higher mathematics, Frege is apparently right. Unified, reasonably clear set-theoretic foundations permit increased confidence in the infinistic methods of modern analysis.[24]) As long as we lack proofs, our mathematical beliefs have the wrong kind of support, even where they are not to be doubted. Our understanding is deficient mathematically, because we lack proofs, and epistemologically, because we are missing a significant avenue to mathematical knowledge.

Frege, of course, *prefers* the logical derivation of arithmetic to any other route – it is not *merely* a significant option. We will later find his attitude reasonable, considerations of certainty aside. The logicist approach is epistemologically more general and illuminating. This defense will align Frege with Kitcher's Dedekind.

Kitcher's main reason for regarding Dedekind as a non-Fregean is Dedekind's vision of evolutionary development, in which mathematicians start from existing mathematics. But Kitcher's contrast is simply wrong. Historicity is consistent with the most rigid apriorism, and, whatever brands of apriorism Frege might endorse, the opening pages of *Die Grundlagen* (1884) stress the continuity with his predecessors' work. Past advances in rigor and systemization encourage the search for foundations from which our present axioms may in turn be derived. Frege's primary motives are, in fact, substantially identical to those Kitcher endorses and attributes to Dedekind. Mathematicians, Frege holds, seek proofs not just to increase certainty but to systemize and unify diverse results. There is, further, the hope that by pushing far enough we may discover previously unrecognized, universally valid principles (e.g. 1884: 2) – just the aim for which Kitcher praises Dedekind. Again, we find agreement where Kitcher tries to force the first-generation logicists apart.

Kitcher portrays Frege as being out of touch with the mathematics of his time, in fact with the logic of mathematical research in general: whereas Frege argues that the developments of (1884) and (1893) arise naturally from mathematics, Kitcher sees a preoccupation with a kind of rigor, and with a Cartesian certainty, that are virtually irrelevant to the working mathematician (e.g. 1984: 268–70; 1986: *passim*). Although Frege's formal advances became useful once paradoxes appeared, he remained, for Kitcher, devoted to a semantical and epistemological programme substantially orthogonal to mathematical practice.

Kitcher is right in two ways. First, Frege pondered strictly philosophical problems early and late. Second, while difficulties in analysis prompted the foundational work of Cauchy, Weierstrass, Cantor, and others, nothing of the kind moved Frege. No questions about the validity of certain methods in the theory of numbers, for example, were to be illuminated by a logical foundation. But to offer these points (particularly the second, which Kitcher stresses) as *charges* is to diminish Frege's genius.

Whatever the philosophical value of logicism, history has upheld its mathematical vision. Mathematics demands proof wherever possible, as Frege and Dedekind observe, and a century ago it became possible to prove even the axioms of arithmetic. The reduction to "logic" does not, and was not intended to, buy certainty. Rather, it shows ordinary arithmetic as a special case of a highly general theory of size and order, one subsuming descriptions of radically diverse structures. This is one of the triumphs of mathematical thought. Of course it was not Frege's alone. But among his contributions are the clear understanding that he was, in a good mathematical sense (and one emphasized by Kitcher!), proving previously unproved propositions. He saw that the demand for such proofs was, precisely, independent of the existence of pressing mathematical difficulties, stemming instead from the drive to generalize and unify. The freedom with which Frege took his leap makes it seem, if anything, particularly inspired.

Further, we now agree that providing set-theoretic foundations, and clarifying their nature as foundations, is impossible without the formal proof theory that Frege invented to that end. So much for "irrelevant rigor."

Frege's mathematical and philosophical efforts point in a single direction: the derivation of arithmetic from logic. It seems clear that the purely rational nature of the premises mattered at least as much as their certainty and, in fact, I think it was primary. If I am (even nearly) right, we can keep a substantially Fregean logicism while rejecting the quest for absolute certainty. Paramount is the relation of arithmetic to the various possible sources of warranted belief.[25]

VIII

Let us briefly address standard worries about analyticity before moving to positive theory.

Quinian concerns are largely irrelevant here. Since logicism does not make set theory true by convention, by definition, or not in virtue of facts, the (fatal) difficulties for such ideas are avoided. And insofar as Quine's point is that individual statements, in mathematics and elsewhere, lack their own meanings (so that questions of their analyticity cannot arise), we can simply deny it.[26] In contrast, his general attack on apriorism is a deep challenge. But we have already considered apriorism. Beyond our earlier remarks, the best response to Quine would perhaps not be abstract epistemological disputation but rather a concrete suggestion as to the character of nonempirical inquiry.

Philosophers of mathematics have fastened on set theory's existence claims as a difficulty for logicism. But recall that purely rational (nonexperiential, nonintuitive) access to objects is possible for the first-generation logicist. An ontology may be analytically justified. Just here, earlier and later conceptions of analyticity diverge. The fundamental point is that if the analyticity thesis concerns the sources of warranted mathematical belief, then the logicist has no reason to hold that mathematics is analytic in the sense of being trivial or consisting purely of definitions. The ontological needs that refute positivist claims of analyticity are therefore no objection at all to Frege and Dedekind. The task is rather to explain *how* we might posit an ontology on nonperceptual grounds; and how this can be rational no matter what. If we even begin to succeed, preemptory dismissals of the idea should carry no weight.

Gödelian incompleteness does defeat a position Frege held. If an idealized being could know all mathematical truths by proving each one in a particular system, then obvious assumptions imply, falsely, that mathematical truth is RE. But this difficulty does not arise for my analyticity thesis. I claim nothing about even potential omniscience. Nor do I assume that our capacity for *a priori* or analytic knowledge of mathematics at any moment is fixed for all time. Reflection on our results and methods may lead us to new principles, hence to previously impossible extensions of our mathematics. In short, logicism makes the epistemological claim that mathematics is (under idealization) developed by pure reason. Any rational faculties employed may be limited and evolving.

Finally, the oft-cited appearance of Russell's paradox shows the logicist only that analytically warranted beliefs are fallible. Since I grant that anyway, there is no further difficulty.

IX

Plato introduced the idea of a systematic, nonempirical discipline in the *Meno*. A request for a definition leads to questions that prompt, step by step, a useful and recognizably correct analysis. This is recollection. Ideally, a structure of such analyses should comprise a unified theory of the objects of platonic inquiry: Justice, Knowledge, Courage, and so forth.

Although the specifics of recollection are tied to Plato's metaphysics and epistemology, the general idea may still be viable. In essence, Plato posits a nonexperiential way (or set of ways) of generating beliefs, plus ways to gain more nonexperiential beliefs from those already given. Any more definite account must assume some particular, controversial theory. Trying to remain somewhat neutral, however, we may say that platonic inquiry proceeds via *reflection* on a body of *conceptual intuitions*. The intuitions may strike us as being obvious, to some degree – as "forcing themselves on us as being true" with more or less insistence. And the intuitive propositions may appear to be, in some sense, part of the concepts they involve.

Reflection may, as in Plato, be a process of asking questions in a way that yields further intuitive beliefs without the aid of experience. It may also be inference. Apriorists might take the inferences to be deductive, but we now know that this restriction is unwarranted. (I believe neither Plato nor Descartes insisted on it.) Thus, considerations on Plato lead quickly to our version of the standard model of mathematical inquiry. Such inquiry is platonic if reflection can dispense with empirical results and if the basic intuitions are, in a suitable sense, strictly conceptual.

Conceptual intuitions as such are in my view not the problem for logicism. The existence of propositions that are in some sense built into our concepts, hence not empirically tested, is now a standard tenet of empirical linguistics. Whatever the difficulties, it is regarded as a matter of linguistic meaning that, if A is longer than B, then B is shorter than A, or that, if John persuaded Mary to leave, then Mary intended to leave. Linguistics is subject to revision, of course. But any element of logicism taken from accepted empirical science will be much better established than any philosophical premises used to challenge it. If the question of logicism comes down to the existence of purely lexical belief, then logicism will be the most firmly grounded position the philosophy of mathematics has ever seen.

Nor, as I remarked, have plausible alternatives to some kind of linguistic or conceptual status for basic beliefs about (say) number been made out. I regret the impossibility of pursuing this (interesting) strand of the argument here; but the attempts to identify perceptual foundations for mathematics are widely understood to be rudimentary and problematic. Indeed, it hardly seems credible that the ultimate evidential basis for set theory should lie in experiences we once had playing with blocks, or that our ancestors had playing with mastodon bones.[27] Quine has long known that this is no way to be a mathematical empiricist.

Thus we may reasonably suppose that the Dedekind–Peano axioms fall out of the concept of number. But here the logicist's real work begins.

We need, first, psychological models that make sense of the notion of conceptual information. Aside from the general obscurity of this issue, the question for logicism is whether any plausible models for terms like "greater than" or "persuaded" also apply to mathematical terms. One reason that it seems nonexplanatory to say that "every number has a successor" is like "brothers are siblings" is that the concepts number and brother may, for all we know, be mentally represented in very different ways. So we need to know what mathematical concepts are before talk of conceptual intuitions in mathematics gains definite content.

Second, standard examples of analyticity lack a crucial feature. They are obvious, while logicism requires analytic propositions that are nontrivial and possibly false. Thus, any model of purely conceptual or lexical meaning taken from linguistics would need to be *generalized* to allow for uncertain conceptual connections. This is a problem cognitive science has hardly addressed.[28] Research has so far focused on a narrow band from a range of cases likely to require a complex theory.

These are not exactly objections. The fact that logicism forces us to confront unsolved problems about mental representation hardly speaks against it. Further, the logicist can now meet a natural challenge.

What we have so far (the objection runs) is really just a brief for the apriority of mathematics, not for a general analyticity thesis. Suppose that mathematics works by nonempirical reflection from fallible conceptual intuitions. Grant also that some model of purely lexical, analytic belief explains how such intuitions can be *a priori*. Now these intuitions need not and do not appear only when mathematicizing begins. Reflection can lead us to consider new concepts, or to consider old ones further, thus generating further intuitions. But even so, we can claim analyticity for the intuitions only. In what good sense is any recondite theorem built into the concept of number as siblinghood is (we allow) linked to brotherhood? Perhaps worse yet, how could the *hypotheses* that, as current wisdom recognizes, pervade mathematics count as *analytic*?

A traditional response is that deduction transfers analyticity from axioms to theorems. Although this may seem a mere stipulation, it makes sense if the point is, as in Frege, to defend the purely rational status of arithmetic against Kantians and empiricists. If we start with purely rational cognition, deduction alone should not lead us outside of that realm. So the present-day logicist may still posit a kind of cognition that is purely rational: that includes as species both conceptual intuition and any reasoning introducing no information gained from other sources. And if this is granted, the nondemonstrative character of some mathematical inferences is no further problem. We are already allowing conceptual intuition itself to be fallible. So analytically justified beliefs can fail. The fallibility of explanatory inference is therefore no bar to the analyticity of the conclusions it yields. If the explananda are themselves analytic, it is reasonable to hold that explanatory inferences preserve analyticity.

But now the logicist runs into difficulties. For mathematics to be analytic in the sense of proceeding from conceptual intuitions is not enough. The logicist also needs a claim of rational necessity; and although that condition, too, is hardly clear, it is obviously still lacking. There are actually two problems here. First, analyticity of the "brothers are siblings" variety is a poor kind of rational necessity. The familiar point is that while anyone with the concept brother may, necessarily, believe that brothers are siblings, having that concept is not rationally necessary at all. Thinking about the family relations induced by sexual reproduction is not a universal principle of reason. Yet for the logicists, arithmetic must follow from universal principles. Garden-variety analyticity is not enough. Second, the necessity of any axioms reached by explanatory inference is problematic. To use hypothesis is to admit, in principle, the existence of viable alternatives. Perhaps one sees none, but one can rarely be sure of their absence. How, then, could one maintain the rational necessity of a given explanation?

LOGICISM

A look at elementary arithmetic confirms these doubts. Many mathematicians and philosophers believe in numbers. Yet others claim (as I did in 1982) that arithmetic is fictional, providing only a convenient, indirect expression of set- or property-theoretic truths. Still other rational beings might discuss cardinality without using the concept of number whose instantiation fictionalists deny. Or so it seems; the issue is further complicated by well-known difficulties about the identities of mathematical concepts (e.g., Benacerraf 1965). (Do Zermelo and von Neumann share a concept of number? Zermelo and Frege? Do Descartes and Archimedes share a concept of curve? etc.) So logicism in one good sense is false after all. Mathematics may be *a priori*, and grounded in conceptual intuition. But it is doubtful whether rationality entails possession of any particular mathematical concept. And clearly, successful theories can be denied on (for example) philosophical grounds, if substitutes are provided. One can, for example, reject set theory in favor of an intensional counterpart. In general, mathematical theorizing is too free for rational necessity to obtain.

But I wish to argue for an attenuated, yet still significant, sense of rational necessity and hence for a still significant logicism. For a change, I will be able to call on Kitcher's support.

Kitcher's Dedekind believes that the mind imposes arithmetical structure on any experienced world. For Kitcher himself, a sanitized, scaled-down version of this idea is an attractive form of logicism (pp. 334–6, 339). Although I have criticized Kitcher, I think he is not entirely wrong. I offer the following position (which is close to Kitcher's at some points, distant elsewhere).

(1) Let us restrict ourselves to theoretical rationality, that is, to rational beings who want to understand as well as survive. For such beings, counting is indispensable in two senses. First, it may be partially constitutive of their rationality. One wonders whether anyone unable to count could be deemed theoretically rational. Second, their theorizing can hardly proceed without data supplied by counting. No matter what world they live in, no matter what they theorize about, they will not get far without raising "how many" questions. In this way, counting is like elementary logical inference. One might call the ability to count a topic-neutral asset.

Counting does not mean having a concept of number. To speak with Frege, counting establishes "statements of number" (*Zahlangaben*), but it involves no numerical ontology, or as Frege put it, no recognition of numbers as objects. Counting is also more elementary than performing arithmetical operations. All we have so far, then, is the rational necessity of primitive arithmetical thought.

(2) Arithmetical calculation also seems rationally necessary, with two qualifications. First, what is necessary is some method of calculation or other. Addition, multiplication, and so forth are ways to find cardinalities of certain sets, given simpler cardinality determinations. They take us from statements of number to statements of number. And it is the statements of number – the answers – that matter, not the particular formalism or theory within which the computation takes place. Second, the necessity of calculating is a step below the necessity of counting. Although a theoretical activity that would benefit from counting but not from further computation is hard to imagine, the degree of rational compulsion drops as one moves from the former to the latter. Agents unable to calculate may still be rational, and they may have some theory, albeit of an impoverished kind.

But allow that *improving* one's theories is also rationally required. For so it appears: insofar as we are theorists, we would be *irrational* to rest content with recognizably limited, incomplete theories. There is a drive to seek theories more comprehensive and simple than whatever we have at any moment. Then computation is rationally necessary also for our impoverished inquirers lacking it, in the sense that the required progress will depend on it. No matter what one's theory, computational ability will lead to improvements.

(3) The pressure to generalize and unify should lead any beings who can calculate to develop a form or analog of number theory. This we may simply assert: in order to understand the arithmetical operations, one needs a theory of a simply infinite sequence and of relations among positions in it. Formulations may vary. One might discuss numbers, sets, infinite sequences in general, or something else. But in one way or another, understanding elementary calculation requires explicit mathematics – the recognition of numbers, or something analogous, as objects.

Thus we have a kind of rational necessity for arithmetic. It is a conditional necessity, since the claim is not that every rational being arithmetizes but that the nature of rational inquiry will compel any rational inquirer eventually to ask arithmetical questions. Nor is any particular style of answer, with its particular ontology and ideology, required. The styles that work presumably share common structure, but any specific one will admit variants.

(4) From here a path to the rational necessity of set theory may open. Roughly, rationality demands a continuing search for explanations (unifications, generalizations), and set theory bears the appropriate relation to arithmetic. But again, qualifications are needed. Set-theoretic explanations will not be unique, since a general theory of cardinality and order may be developed in other terms; moreover, significant alternatives exist within set theory itself. On this matter, however, our remarks about arithmetic hold *mutatis mutandis*. It may be rationally necessary to generalize in the set-theoretic manner, whatever specific theory one chooses. Even so, reflection on arithmetic can take us only so far. A sequence of steps, leading from weaker to stronger set theories by iterated plausible generalization, will hardly force on us such axioms as replacement and (unrestricted) choice.

Beginning with more than arithmetic, say with analysis, still does not lead to full set theory. So a case for the rational necessity, in any sense, of any strong set theory would be likely to appeal not just to the considerations of (1)–(3) but to the naturalness and explanatory power of progressively stronger principles of comprehension or infinity. I think this is not an unreasonable expectation (e.g. Maddy 1988). That, however, certainly needs detailed study.

One might experiment with a different route to full set theory. Suppose that some degree of rational necessity were demonstrated for a great variety of instances of an unrestricted comprehension principle. The idea, very roughly, would be to show that a theorist is always better for being able to speak of the set of all tomatoes, of prime numbers, of Frege's ancestors, and so forth – that any thinker anywhere gains by positing extensions (or intensions) for predicates. The truth of naive comprehension would then present itself as an explanatory hypothesis. With the discovery of Russell's paradox, one would be rationally compelled to seek an alternative – a minimally restrictive one, in order to keep as much power as possible. This would lead to something substantially like current higher set theory. I hope to develop this idea elsewhere. The attempt to reach a kind of analyticity for set theory from the dual starting points of arithmetical thought and run-of-the-mill abstraction has an obvious historical appeal.

To summarize, any rational inquiry needs counting. If it needs counting, then calculation will advance it further. The required search for progressively better explanations will, however, drive a calculating inquirer to pure mathematics: first to something like arithmetic, then to set theory. Mathematics at each stage is a collection of theorems deduced from axioms, where the axioms rest on a combination of (nonempirical) explanatory power and one's grasp of the concepts that inquiry forced one to develop – such as number and set. This I submit as a form of logicism.

Logicism could hold only under idealization, since real people need blackboards and computers. Also, we have been assuming inquirers free from practical demands; otherwise, needs for food and shelter might override the requirement to seek better explanations. But this will have been understood. What is interesting is whether logicism is true even for ideal reasoners. And this comes down to issues about rationality. If sophisticated rational inquiry could proceed (perhaps in another world, or for psychologically different beings) without addressing cardinality questions, then our claim about counting would be false. Failing that, suppose that some style of explanation, radically different from any now used, were acceptably to subsume the facts of counting (and calculation) without introducing anything like a number theory. It would follow that arithmetic is not, even in my special sense, logical. (Here the logicist might retrench, proposing that arithmetic is logical relative to certain general conditions on what explanation is. That would be a further attenuation.) Similarly, if one could rationally subsume arithmetic under something very different from set theory, then set theory would not be logical.

That logicism should ultimately turn on these questions is just right. Frege and Dedekind held counting and abstraction to be fundamental to any rational thought whatsoever. If that is false, logicism should be false. But these are now fantasies. We can imagine no serious inquiry without counting. We see no way to understand the facts of counting except through recognizable mathematics. Within mathematics, something like, or implicitly containing, a theory of sets seems necessary for the perspicuous discussion of cardinality. *If* we are right in all this, and *if* mathematical inquiry is independent of perception to the extent here conjectured, then I would say that a rationalist mathematical epistemology is true. In other words, mathematics is logic.

For Paul Benacerraf

LOGICISM

NOTES

1. These remarks were stimulated by conversations and correspondence with Stuart Shapiro, who discusses my 1987 paper in forthcoming work. Shapiro has also said that my position favors proof-theoretic over model-theoretic views of logical consequence. I disagree, since any view of the latter kind must somehow specify logical constants, which may be done with reference to their epistemic properties in a way I could accept (cf. McCarthy 1987). One shortcoming of my 1987 essay is that the fundamentality condition (pp. 9-10) is too briefly motivated; that aside, there is a careless slip about the negation of CH (p. 5).
2. Although the differences between set-theoretic and other forms of logicism are sometimes important, I will usually ignore fine points and speak of logicism as reducing arithmetic to set theory. It is generally agreed that any alternative foundation would have to be strong enough to interpret Zermelo-Fraenkel set theory.
3. Charles Parsons's subtle attempts to find a Kantian theory of intuition (e.g. 1979-80) are exempt from my quick dismissal here. I hope to discuss them elsewhere.
4. For general background on this historical situation, see especially Sluga (1980); also Kitcher (1979). I discuss some aspects of Frege's Kantianism about numbers and sets in detail in my 1983a paper. The special features of first-generation logicism are emphasized there, along with brief remarks on Dedekind's kinship with Frege.
5. Wagner 1983a develops this idea. This reading does not (*pace*, for example, Schirn (1988: 999) conflict with Frege's insistence that thoughts (judgeable contents) are not the products of our mental processes. I take that insistence to refer to *empirical* mental processes, in the context of a dispute with naturalistic and psychologistic opposition. It is as though a Kantian were to say that tables are independent of our thinking; as this is meant, it leaves open the transcendentally ideal character of bodies. Similarly, my view is that thinking in the ordinary sense depends on prior structuring processes that yield the judgeable contents. Extensions, hence numbers, are among their products (see Wagner 1983a).
6. Citations will henceforth be to Kitcher (1986) unless otherwise indicated.
7. For the themes of this paragraph, see Kitcher (1984: chs 1-5).
8. See further the discussion of APW and PPT apriority below.
9. Dedekind's admirers seem to have trouble taking his constructivism seriously – e.g. Stein (1988: 249), who glosses Dedekind's talk of free creation as an expression of general toleration and open-mindedness.
10. Even if it works. Why suppose that you, I, Newton, and Kitcher all have the same idealized operations of computing and collecting in mind?
11. Nor does Kitcher even have any real argument for his further claim that Dedekind and Frege differed over the possibility of tacit foundational beliefs. That issue, however, is tangential here.

12 An extensive interdisciplinary literature addresses these questions. See Cummins (1989) for one recent bibliography.
13 See pp. 28–31 for more on the issues of the next few paragraphs. Wagner (1983b) approaches the same position – that some main rationalist ideas are true but require idealized or communal knowers – from another direction.
14 Otherwise, they are epistemologically like us. Their nonexperiential sources may be as fallible as ours, and they may have no great ability to find axioms that would settle outstanding mathematical questions.
15 My forthcoming paper discusses, in a different context, some of the subtle problems this attitude involves.
16 "On the rival picture, the history of mathematics is punctuated by events in which individuals are illuminated by new insights *that bear no particular relation to the antecedent state of the discipline*" (Kitcher 1988: 297; his emphasis).
17 "[Zermelo's] justification is exactly analogous to that of a scientist who introduces a novel collection of theoretical principles on the grounds that they can explain the results achieved by previous workers in the field" (Kitcher 1988: 295).
18 The claim that mathematical justification is hypothetico-deductive is primarily Quine's (1951, 1960). But for Quine, mathematics is simply more empirical science. Main defenses of intrinsically mathematical explanation include Lakatos (1976), Steiner (1978), and Kitcher (1984).
19 This may lie behind Goldfarb's doubts about Kitcher (1988: 70, 80).
20 I assume that the argument from quantum logic, in particular, is too obscure and controversial to count.
21 This qualification will now often be left implicit.
22 Remember that we are idealizing. For example, a number thought prime on *a priori* grounds could be factored by a computer – but someone with enough memory and time could obtain the same refutation *a priori.*
23 Consider Maddy (1988) on the axiom of measurable cardinals. In the presence of *massive* hypothetico-deductive support, deduction from this axiom is treated by *some* set theorists as *tantamount* to proof (p. 508)!
24 Compare natural science, where a systematic explanation can greatly increase one's confidence in the data, even as they support it.
25 The close of Stein (1988): "As to 'logicism', as it seems to me the vaguest of the three doctrines, I find it hardest to envisage prospects for it, and am most strongly inclined to see its positive contribution as exhausted.... Nevertheless, it is (barely, I think) conceivable that progress in understanding the structure of 'knowledge' will succeed in isolating a special kind of knowledge that reasonably deserves to be considered 'logical', thus conferring new interesting content upon the question whether mathematics does or does not belong to logic" (pp. 258–9). Stein has the problem exactly right, although I obviously disagree about the prospects.
26 My 1986 paper gives one possible analysis of Quine's missteps (Eric

Koski has convinced me that it cannot be the whole story).
27 Frege (1884: VII–VIII) is still pertinent: "What, then, are we to say of those who ... betake themselves to the nursery, or bury themselves in the remotest conceivable periods of human evolution, there to discover ... some gingerbread or pebble arithmetic! ... It is as though everyone who wished to know about America were to try to put himself back in the position of Columbus, at the time when he caught the first dubious glimpse of his supposed India."
28 I think these connections must clearly be rather different from the familiar paradigm-based defeasible inferences (e.g. that birds fly, that Republicans aren't pacifists), and hence that they call for quite different models. I hope to develop this point elsewhere.

REFERENCES

Aspray, W. and Kitcher, P. (eds) (1988) *History and Philosophy of Modern Mathematics*, Minneapolis, MN: University of Minnesota Press.

Benacerraf, P. (1965) "What numbers could not be," *Philosophical Review* 74: 47–73. Reprinted in P. Benacerraf and H. Putnam (eds) *Philosophy of Mathematics*, 2nd edn, Cambridge: Cambridge University Press, 1983.

Benacerraf, P. and Putnam, H. (eds) (1983) *Philosophy of Mathematics*. 2nd edn, Cambridge: Cambridge University Press.

Cummins, R. (1989) *Meaning and Mental Representation*, Cambridge, MA: MIT Press.

Dedekind, R. (1888) *Was Sind und was Sollen die Zahlen?*, Braunschweig: Vieweg; reprinted 1969. English translation in R. Dedekind *Dedekind's Essays on the Theory of Numbers*, New York: Dover Publications, 1963.

—— (1963) *Dedekind's Essays on the Theory of Numbers*, New York: Dover Publications.

Foley, R. (1987) *The Theory of Epistemic Rationality*, Cambridge, MA: Harvard University Press.

Frege, G. (1884) *Die Grundlagen der Arithmetik*, Breslau: Koebner. English translation in G. Frege *The Foundations of Arithmetic*, Evanston, IL: Northwestern University Press, 1974.

—— (1893) *Die Grundgesetze der Arithmetik*, vol. 1, Jena: Pohle; reprinted Hildesheim: Olms, 1962. Partial translations in G. Frege *Translations from the Philosophical Writings of Gottlob Frege*, Oxford: Basil Blackwell, 1970, and *The Basic Laws of Arithmetic*, Berkeley, CA: University of California Press, 1976.

—— (1895) "Critical notice of E. Schröder's *Vorlesungen über die Algebra der Logik*," reprinted in G. Frege *Kleine Schriften*, Darmstadt: Wissenschaftliche Buchgesellschaft, 1967; translated in G. Frege *Translations from the Philosophical Writings of Gottlob Frege*, Oxford: Basil Blackwell, 1970.

—— (1903) *Die Grundgesetze der Arithmetik*, vol. 2, Jena: Pohle;

reprinted Hildesheim: Olms, 1962. Partial translation in G. Frege *Translations from the Philosophical Writings of Gottlob Frege*, Oxford: Basil Blackwell, 1970.

—— (1967) *Kleine Schriften*, Darmstadt: Wissenschaftliche Buchgesellschaft.

—— (1970) *Translations from the Philosophical Writings of Gottlob Frege*, Oxford: Basil Blackwell.

—— (1974) *The Foundations of Arithmetic*, Evanston, IL: Northwestern University Press.

—— (1976) *The Basic Laws of Arithmetic*, Berkeley, CA: University of California Press.

—— (1986) *Die Grundlagen der Arithmetik* (critical edition of 1884 edition with supplementary texts), Hamburg: Felix Meiner.

Friedman, M. (1987) "Carnap's *Aufbau* revisited," *Nous* 21: 521–45.

Gödel, K. (1947/64) "What is Cantor's continuum problem?," reprinted in P. Benacerraf and H. Putnam (eds) *Philosophy of Mathematics*, 2nd edn, Cambridge: Cambridge University Press, 1983.

Goldfarb, W. (1988) "Poincaré against the logicists," in W. Aspray and P. Kitcher (eds) *History and Philosophy of Modern Mathematics*, Minneapolis, MN: University of Minnesota Press.

Haaparanta, L. and Hintikka, J. (eds) (1986) *Frege Synthesized*, Dordrecht: D. Reidel.

Kitcher, P. (1979) "Frege's epistemology," *Philosophical Review* 88: 235–62.

—— (1984) *The Nature of Mathematical Knowledge*, Oxford: Oxford University Press.

—— (1986) "Frege, Dedekind, and the philosophy of mathematics," in L. Haaparanta and J. Hintikka (eds) *Frege Synthesized*, Dordrecht: D. Reidel.

—— (1988) "Mathematical naturalism," in W. Aspray and P. Kitcher (eds) *History and Philosophy of Modern Mathematics*, Minneapolis, MN: University of Minnesota Press.

Lakatos, I. (1976) *Proofs and Refutations*, Cambridge: Cambridge University Press.

McCarthy, T. (1987) "Modality, invariance, and logical truth," *Journal of Philosophical Logic* 16: 423–43.

Maddy, P. (1980) "Perception and mathematical intuition," *Philosophical Review* 89: 163–96.

—— (1988) "Believing the axioms," *Journal of Symbolic Logic* 53: 481–511, 736–63.

Parsons, C. (1979–80) "Mathematical intuition," *Proceedings of the Aristotelian Society N.S.* 80: 142–68.

Putnam, H. (1975) "What is mathematical truth?," in *Mathematics, Matter and Method*, Cambridge: Cambridge University Press.

Quine, W.V. (1951) "Two dogmas of empiricism," *Philosophical Review* 60. Reprinted in W.V. Quine *From a Logical Point of View*, 2nd edn, revised, Cambridge, MA: Harvard University Press, 1980.

—— (1960) *Word and Object*, Cambridge, MA: MIT Press.

—— (1980) *From a Logical Point of View*, 2nd edn, revised, Cambridge,

MA: Harvard University Press.
Robinson, H. (ed.) (forthcoming) *Problems for Physicalism*, Oxford: Oxford University Press.
Russell, B. and Whitehead, A.N. (1910) *Principia Mathematica*, vol. 1, Cambridge: Cambridge University Press.
Schirn, M. (1988) "Review of Frege *Die Grundlagen der Arithmetik*, 1986 edition," *Journal of Symbolic Logic* 53: 993–9.
Sluga, H. (1980) *Gottlob Frege*, London and Boston, MA: Routledge & Kegan Paul.
Stein, H. (1988) "*Logos*, Logic, and *Logistike*," in W. Aspray and P. Kitcher (eds) *History and Philosophy of Modern Mathematics*, Minneapolis, MN: University of Minnesota Press.
Steiner, M. (1978) "Mathematical explanation," *Philosophical Studies* 37: 135–51.
Wagner, S. (1982) "Arithmetical fiction," *Pacific Philosophical Quarterly* 63: 255–69.
—— (1983a), "Frege's definition of number," *Notre Dame Journal of Formal Logic* 24: 1–21.
—— (1983b) "Cartesian epistemology," unpublished manuscript.
—— (1986) "Quine's holism," *Analysis* 46: 1–6.
—— (1987) "The rationalist conception of logic," *Notre Dame Journal of Formal Logic* 28: 3–35.
—— (forthcoming) "Truth, physicalism and ultimate theory," in H. Robinson (ed.) *Problems for Physicalism*, Oxford: Oxford University Press.

4

EMPIRICAL INQUIRY AND PROOF

Shelley Stillwell

SUMMARY

In this paper the later Wittgenstein's distinction between proofs and experiments and its place in his philosophy of mathematics is explored. Section I observes that his apriorism must also be considered in this context, for he makes a parallel division between the sorts of statements which proofs or experiments establish. Sections II and III provide, respectively, background material about the proof/experiment distinction and the rather persuasive evidence Wittgenstein gives for drawing it. There is some discussion of his thesis that proofs are models or rules governing empirical investigation, but lengthier comments about certain fundamental, "logical" features he ascribes to proofs.

I defend Shanker's suggestion that such logical features are more important to our understanding of Wittgenstein's philosophy of mathematics than an epistemic property such as perspicuity. In Section IV, I explain the important difference between logical features of proof, on the one hand, and perspicuity, on the other. I then review a substantial attack on apriorism, an attack from which Wittgenstein is defended in Section V. Key to that defense is his claim that proofs are proof-*types*, not *token*-constructions. Thus, on my account of the matter, the knowledge of proofs involves belief pertaining to construction-types rather than specific constructions.

Section V uses this account to argue that our knowledge of proofs, not only does not include, but also is significantly independent of, conviction about token-constructions. While Wittgenstein is certainly no platonist, my discussion hopefully exhibits how his notions of mathematical proof and mathematical knowledge are, so to speak, less dependent on contingent states of affairs than is standardly thought.

I. INTRODUCTION

In an early review of *Remarks on the Foundations of Mathematics*,[1] Kreisel approved a key objection Wittgenstein makes against "naive" empiricism. The naive empiricist assumes "a sharp distinction between the empirically given and the means of description," whereas Wittgenstein contends: "we need concepts to tell us what are the facts."[2] His famous claim is that whatever sensory experience may seem itself to teach us must be shaped as well by features of human conceptualization. Now a distinction between proofs and experiments and, more broadly, between mathematics and experience is a focal point in Wittgenstein's later philosophy of mathematics. This does not reinstate the illicit contrast between concepts and data. Wittgenstein again is concerned to maintain that mathematical conceptualization shapes certain experiences, and he insists that *all* our concepts are to a degree dependent on "very general facts of nature."[3]

Our goal is to understand the significance of Wittgenstein's proof/experiment distinction in his philosophy of mathematics. Section II sketches background material, and Section III examines his evidence for the distinction. I find the evidence reasonable and persuasive. It also clarifies why Wittgenstein surprises readers with the apparently doctrinaire thesis that proof must be "surveyable." An additional, related distinction is in need of clarification, namely that between types of statements which proof and experimentation establish. For Wittgenstein, experimentally confirmed statements are descriptive of perceptual experience. The results of proof, on the other hand, constitute *a priori* standards of correctness for carrying out certain empirical procedures. Here Wittgenstein has returned to something at least resembling traditional apriorism and requiring interpretation. In Sections IV and V, we try on his behalf to deflate one important attack on apriorism.

II. PROOFS AND EXPERIMENTS: THE DISTINCTION

For Wittgenstein, experiments are describable in entirely empirical terms. In principle at least,[4] one can comprehend an experiment through descriptions – empirical propositions – which have clear meaning prior to the conduct of the experiment. Proofs characteristically introduce new concepts, and so possess content the recording of which goes beyond descriptive language. Grasping a proof involves seeing its conceptually creative significance. Proof, then, cannot just be one form of experiment.

A proof, such as calculation of an equation, is "the picture of an experiment as it practically always turns out" (1956: VII, 18; cf. III, 23). Crucially, "this picture stands in need of ratification, and that we give it when we work over it" (VII, 9); for example, we check computations someone has previously performed. If proof is to be possible, it must be possible for our results to be generally "ratified" by others. Wittgenstein says, "The agreement of ratifications is the pre-condition of our language-game ...," here the "game" of calculating products, sums, and so on for a variety of purposes. (How to individuate the "games" is not our concern.)

To illustrate, most of us "agree" in obtaining "148" when we try to calculate the product of 74 and 2. But the mathematical statement "$74 \times 2 = 148$" does not assert this empirical fact (cf. VI, 23, 26, 30). As Wittgenstein puts it, the "agreement" in what results we generally obtain "is not affirmed in [the game]" (VII, 9). While agreement-in-practice underwrites the multitude of demonstrative "games" we play, *mathematical statements* of results do not describe anything merely empirical.[5] "Surely it remains an empirical fact that men calculate like this!"; "but that does not make the propositions used in calculating into empirical propositions" (VII, 18).

Wittgenstein provides a useful analogy:

> If someone asks me: "What colour is this book?" and I reply: "It's green" ... might I as well have given the answer: "The generality of English-speaking people call that 'green'"?
>
> Might he not ask: "And what colour do *you* call it?" For he wanted to get my reaction.
>
> (III, 71)

The point is clear. Finding out what the product of 74×2 is is not finding out what people "believe" the product is. On a straightforward construal, the request "What is 74×2?" seeks information about *what* one believes, and is not a question about belief *per se*. Mathematical propositions do not describe epistemic states, nor correlations among these states which may show up in our linguistic practice.

This point recurs in discussions of the general import or significance of finding *proof*: "We say, not: 'So *that's* how we go!', but: 'So *that's* how it goes!'" (III, 69).[6] And "It is not our getting this result, but its being the end of this route, that makes us accept it" (III, 39). Its being, as it seems to us, the result is in a sense the "reason" we accept it. The result is "the end of this route [the proof]" (III, 39). ("The proof is the route taken by the test" (VI, 2). See I, 163.) Importantly, Wittgenstein insists that this or that *being* the correct result, as opposed to someone's *getting* a result, is not temporally relative. Recall the "creative" aspect of proof:

> The proof is now our model of correctly counting 200 apples and 200 apples together: that is to say, it defines a new concept: "the counting of 200 and 200 objects together". Or, as we could also say: "a new criterion for nothing's having been lost or added".
>
> (III, 24)

Shortly before this we find: "'This is the model for the addition of 200 and 200' – not: 'this is the model of the fact that 200 and 200 added together yield 400'. The process of adding *did* indeed yield 400, but now we take this result as the criterion for the correct addition – or simply: for the addition – of these numbers." The contention, obviously, is that the use of tenses is incorrect in statements expressing mathematical results.

Elsewhere, Wittgenstein's interlocuter queries, "If the calculation has been done right, then this must be the result. Must *this always* be the result, in that case?" "Of course," Wittgenstein responds (IV, 35). What Wittgenstein means is that mathematical truths *timelessly* express what they do (see VI, 36). (For exemplary cases, also examine I, 23, 103; VI, 2; and VII, 69.)

To return to the issue, we said that Wittgenstein excludes empirical description from the content of mathematical results (or truths). The same holds regarding the substance or content of entire proofs. While definitive passages are rarely available, some of Wittgenstein's remarks are suggestive. For example, he says:

> The limit of the empirical – is *concept-formation.*
>
> What is the transition that I make [in a proof] from "It will be like this" to "it *must* be like this"? I form a different concept....
>
> But what if someone now says: "I am not aware of these *two* processes, I am only aware of the empirical, not of a formation and transformation of concepts which is independent of it...."
>
> So much is clear: when someone says: "If you follow the *rule*, it *must* be like this", he has not any *clear* concept of what experience would correspond to the opposite.
>
> (IV, 29)

The perceptual reports favored by empiricist philosophers of mathematics, report "the opposite" of which is conceivable, play no semantic role in mathematical demonstration. The "formation and transformation of concepts" typifying proof-construction involve content which cannot be described by describing our observations or empirical processes.

Wittgenstein continually attacks the extreme empiricism we associate with Mill, especially the reduction of mathematical procedures to empirical ones.

> It could be said: a proof helps communication.
> An experiment presupposes it.
> Or even: a mathematical proof moulds our language.
> ...
> *The limits of empiricism.*
>
> (III, 71)

A thoroughgoing empiricism cannot account for what Wittgenstein sees as special aspects of mathematical proof-construction, nor explain contrasts between these and some features of empirical inquiry. Given the seemingly widespread impression that, with the logical empiricist, Wittgenstein believes mathematics is analytic or "definitional," one might wonder whether logical empiricism fares better in this context. Unlike this sort of empiricist, though, the sense in which Wittgenstein uses "definitional" does not exclude being *synthetic* (clear endorsements of the doctrine of the synthetic *a priori* occur in IV, 39, 43, and VI, 36).

Demonstration or proof, we said, yields standards. A proof "models" what it is like to do certain things correctly, therein imposing some limitations on permissible descriptions of experience. Our primary means of empirical inquiry, namely, observation and experiment, cannot set standards or serve as models in the same way. Instead, empirical inquiries obey the mathematical requirements "created" by proof. An example may be useful:

> The proof is our new model for what it is like if nothing gets added and nothing taken away when we count correctly etc.
>
> (III, 39)

Counting and "adding up" groups of apples are activities governed by arithmetic, specifically standards of addition. The standards are not governed by the procedure or "experience" of counting and "adding" the apples. This is a commonplace in the literature, and it casts light on Wittgenstein's reference to "the role of miscalculation.... The role of the proposition: 'I must have miscalculated'. It is really the key to an understanding of the 'foundations' of mathematics" (III, 90).

Consider this passage:

> the proposition 25 × 25 = 625 ... is so to speak an empirical proposition hardened into a rule. It stipulates that the rule has been followed only when that is the result of the multiplication. *It is thus withdrawn from being checked by experience, but now serves as a paradigm for judging experience.*
> (VI, 23: italics added)

A standard is "set up" in the course of the proof, which thereby yields a method for judging some experience. Its precise content may be elusive, but the point of the standard is to regulate: "mathematics is *normative*" (VII, 61). Since a proof in a way aims, then, at establishing a "regulation," the content of the proof must not appeal to experiential data. For, in that event, deployment of the standard would be contingent on confirmation of an empirical matter, and this would destroy the *unconditional* nature of mathematical standards, that is, their necessity (not to be confused with the "un-Wittgensteinian" idea of standards that are wholly *unconditioned*). Appeal to experience also would undercut Wittgenstein's thesis that mathematical truths legislate via concepts which neither derive from experience nor depend on it for the correctness of their specific applications. So empirical propositions cannot be essential to the content of proofs (see VI, 28, 46), and neither, in particular, can descriptions of experimental processes.

III. THE PROOF/EXPERIMENT DISTINCTION: SUPPORTING EVIDENCE

Wittgenstein remarks: "'Proof must be capable of being taken in' [i.e. must be surveyable] really means nothing but: a proof is not an experiment" (III, 39). This is surely no analysis of "surveyable." Rather, it reflects three points: (i) there are essential differences between the concepts *proof* and *experiment*; (ii) these differences are detectable in contrasting aspects or these concepts (e.g. the contrast between "*repeating* a proof" and "*repeating* an experiment," discussed below); finally, the general point of the passage is that (iii) surveyability "means" – that is, consists of – the group of features which accrue to proofs in virtue of those contrasts cited in (ii).

I want to follow the lead of Stuart Shanker[7] who suggests that Wittgenstein's "surveyability" refers mainly to logical rather than epistemic features of proof. The epistemic construal is the standard one: "surveyability" is viewed as synonymous with "perspicuity," which refers to a characteristic clarity of proofs making them knowable.[8] But, as Shanker says, Wittgenstein often tries to clarify the notion of surveyability by gesturing towards whatever there is in proofs that drives a conceptual wedge between them and experiments.[9] The quotation with which this section opens provides an excellent example.

We also find Wittgenstein using "surveyable" (*übersehbar*) to stand for some more specific aspect of the proof/experiment distinction. Thus he makes the claim that the statement, "Proofs must be surveyable," "aims at drawing our attention to the difference between the concepts of 'repeating a proof', and 'repeating an experiment'" (III, 55). And he notes:

> To repeat a proof means, not to reproduce the conditions under which a particular result was once obtained, but to repeat every step *and the result*. And ... this shows that proof is something that must be capable of being reproduced *in toto* automatically, ...
>
> (III, 55)

(We ignore the reference to "automatic" procedures. What Wittgenstein has in mind never becomes clear, and he says little about this idea.) The major point is that repetition of a proof is "holistic" in that "repeating the proof" entails *reproducing its results*. Indeed, we may call proofs themselves "holistic," since their results are part of their very identity. (That Wittgenstein sees it this way too comes out in III, 44.)

On the other hand, genuine reconstruction of experiments entails openness concerning what the result will be. Otherwise, the fresh experiment is a bogus "repetition" of the old, because the outcome has been "rigged" (cf. I, 109). This constraint makes meaningful a common purpose of repeating experiments, namely, to gather evidence for or against previous results of the same experiment. A proof is not repeated to gain such evidence but, for instance, to show others how it "goes," to gain deeper appreciation of strategy, to enjoy its details, or the like.

Admittedly, one can "check the result" of a construction one believes, yet is not certain, is a proof. One possible outcome is that one proves that the correct result (R) is not what one had thought. Here one makes a construction *proving R*, the earlier construction proving nothing, or perhaps proving something irrelevant. Clearly, the checking activity may or may not yield information helping one to assess that first construction. To see that the result one sought should rather be R is not necessarily to discover anything about one's earlier constructive efforts. Contrast this with the experimental case, where the outcome of checking must give evidence about earlier experimentation. If one finds that fixed experimental conditions first had one result and later another, this is evidence against this type of experiment's establishing either result. (Under the experimental conditions, the earlier result was, say, that such and such occurs, but it is found that under the fixed conditions something else may happen.)

One could, of course, have instead found the conditions of the experiment to lead to the same result in different cases. Then, however, the usefulness of the experiment is confirmed, that is, some kind of connection between its conditions and the mentioned result is supported. But consider the case where checking up on the result of what seemed to be a "proof" shows one that the result R of that construction *is* the (correct) result. Now Wittgenstein's view is that, prior to one's present finding about the correct result, one's position was epistemically defective (cf. App. II, 1–8, and VII, 18). During the checking activity one makes a new construction, and this assists one in grasping that *this type* of construction proves R. Yet grasping this fact need not confirm that the "old" construction proves R. We can imagine that one does see the old construction, like the new one, to be of the type which, as one just found, establishes R. Still, present constructive work is not direct support for the sameness of the two constructive efforts. At most, then, "deriving" from a given construction a clearer *proof-*construction will show that, *if* the first construction is of the relevant type, then it establishes R. (Section V deals with pertinent type-token issues.)

Now observe a second difference between proofs and experiments, namely, a basic respect in which experiments, yet not proofs, are "conditioned." As the identity of a proof includes its result, so that of an experiment includes the experimental conditions under which it is carried out. This is why "repeating an experiment" entails reproducing some of the "conditions under which a particular result was once obtained" (III, 55 above). By contrast, "repeating a proof" does not entail reproducing any particular empirical circumstances.[10]

Wittgenstein's reasoning seems unimpeachable: The concepts *repeating a proof* and *repeating an experiment* are distinct; for instantiating the first requires reproducing the proof's result and does not require replicating specific experiential conditions, while instantiating the second requires reproducing the experimental conditions but precludes "rigging" the experimental outcome; it follows by a good Fregean principle that the concepts *proof* and *experiment*, or the senses of "proof" and of "experiment," are distinct, that is, different.

This much may strike the empiricist as superficial. She/he may hold that the interesting epistemological issue is whether specific proof-constructions are *entirely* distinct from all experiments. The point against apriorism may be, not that there is *no* difference between experimental and proof procedures, but that full-scale demonstrations must be "empirical" in virtue of including some experimental components. We will see that Wittgenstein allows specific proof-constructions and experimental processes to "coincide and, in a sense, allows that proof-construction involves experimenting." But he maintains that neither concession undermines the distinction between proofs and experiments. His distinction goes much deeper than our discussion has so far revealed.

A third contrast Wittgenstein draws between proof and experimentation is this: "I can calculate in the medium of imagination, but not experiment" (I, 98). He holds that in constructing a certain proof one "could ... have *imagined*" the "characteristic positions and movements of" the proof-picture's components, in this case the links of a chain (I, 79–80). In contrast, however,

> you can't get to know any property of the material [the chain of links] by imagining. The experimental process [of manipulating the chain to see what shapes result] disappears when one looks at the process simply as a *memorable picture* [of "the fact that this can be done with it"].
>
> (I, 80; italics added)

The claim is that in empirical inquiry imagination cannot take the place of "running" experiments, or for that matter of single observations in space at a time. With proofs, though, imagination may always substitute for, for example, marks on paper.

I am not certain that Wittgenstein is right about the role of imagination. It is beyond the scope of this essay to assess his claim, but nonetheless we need to guard against misinterpreting it. The "reliability" of imagination for purposes of proof-construction should not be taken to refer to a specifiable imaginative process, as if a definite kind of experience were to account for our knowledge of proofs. The idea of "inner" experience leading to knowledge is in tension with central tenets of Wittgenstein's philosophy of mind and epistemology. Besides, he says, "Certainly, experience tells me how the calculation comes out [on a specific occasion]; but that is not all there is to my accepting it" (I, 164). What leads me to accept "my" result as the *result* – what "tells us it" – is "*imagination*," "And the germ of truth is here; only one must understand it right" (IV, 4). Even more explicitly, Wittgenstein asserts that a mathematical result "is not designed to express any *experience*, but rather to express the *impossibility* of imagining anything different" (IV, 6; italics added).

What any of us imagines is just as contingent and "unmathematical" as what any of us obtains as the result of a construction. The connection between our grasp of proofs and our imagination rather concerns "the *impossibility* of imagining anything different." Another passage early in IV is useful in this context: "It is not our finding the proposition self-evidently true, but our *making* the self-evidence *count*, that makes it into a *mathematical* proposition" (IV, 3, italics added). Replace the "self-evident proposition" with a "proof-picture" which proves it and is presented to imagination as something the opposite of which is unimaginable. Then "seeing" that nothing else is imaginable is comparable with being "struck" by the self-evidence of propositions. Thus, as IV, 3, says that the latter kind of experience cannot establish a mathematical proposition, so "imaginative seeing" cannot establish a mathematical result, that is, give us knowledge that a construction is a proof.

Let us turn to Wittgenstein's interesting concession that proof-constructions may not only be part of, but even *be* experiments. It is "*obvious*," he tells us, that a calculation may be "a part of the technique of an experiment ..." without thereby being an experiment itself (VII, 17). Clearly, this is right since the use of mathematics in constructing and running experiments is hardly evidence for empiricism.[11] When Wittgenstein grants that "the action of calculating can also *be* an experiment," he says that this depends on the "point of view" from which the action of calculating is seen (VII, 17, italics added). Examples arise if we want "to test what this man calculates, in such-and-such circumstances, when set this question." We could then experiment to see whether the man will obtain the correct result, or how long it will take him to find "what 52 × 63 is," and so forth (VII, 17).

Applicable commentary about such cases occurs in III, 67–9. There Wittgenstein argues that, even if a calculation is an experiment in some situation, a crucial difference between the calculation and the experiment remains. (Thus this specific calculating activity and this experiment are not strictly identical in the sense of having precisely the same things true of them.) Wittgenstein claims that what can be considered a "mistake in calculating" may well differ from a "*mistake* in the experiment," and a mistake in the calculation may be recognizable, not as a mistake in but as a "*miscarriage* of the experiment" (III, 68).

To support this distinction, he notes that a mistake in an experiment occurs when one fails to "observe ... the *conditions* of the experiment." Assume we want to see whether Dee, an adult with average educational background, can calculate well at a particular level of complexity. Presumably, a condition of our experiment would be that Dee perform her tasks in relatively quiet surroundings. Imagine we violated this condition and "made [Dee] calculate when a terrible noise was going on" (III, 68). Intuitively, this would constitute a "mistake" in the experiment itself, specifically an error in the implementation of its (presumed) design. Yet, Wittgenstein observes, from the occurrence of such a mistake we can infer nothing about the quality of Dee's performance. Perhaps she will be unable to calculate at all, or will calculate, only with countless errors. *Or* she might flawlessly perform all calculations put to her. Thus an experiment E can be an activity of calculating C, yet a *mistake in E* need not be the same as a *mistake in C.*

On the other hand, if Dee were to make mistakes, this fact would yield no information about the experimental conditions surrounding her assigned tasks. Dee or anyone else might miscalculate, with or without terrible noise going on. Hence, the fact that she makes errors does not imply that the conditions of the experiment have been violated. Now return to the case where there is flawless calculation despite the presence of noise. This case shows that there can be (correct) *proof*-constructions, that is, actions of constructing what "embody" proofs, which are *incorrect experiments*. Therefore, the concepts *proof* and *experiment* differ also by reason of the fact that their subconcepts (*correct*) *proof-construction* and *correct experiment* are distinct. The earlier contrasts Wittgenstein drew, for example, between repeating proofs and repeating experiments, will also apply in the present example. What it is to repeat the proofs Dee constructs can easily diverge from what occurs in repeating her experimental activity. As before, repeating her proofs necessarily requires the new proofs to include the same results as the old. That constraint is inapplicable for purposes of repeating the experiment, since Dee's new constructions may be incorrect calculations even if the corresponding constructions were not.

Wittgenstein has at the least shown that proof cannot be analyzed as a special kind of experimentation. Additionally, he seems right to hold that, even if sometimes specific calculatory and experimental activities are the same, the former may be correct calculation while the latter is improper experimentation, or the calculation may contain errors while the experiment does not. One could even feel inclined to defend the stronger view that *no* instance of experimental activity can be strictly identical to *any* instance of calculatory activity. This would require some general argument to the effect, say, that even if E is C whatever *could* be a *mistake* in E *could not* be a *mistake* in C. Unfortunately, Wittgenstein does not attempt any general argument along these lines.

It is worth pausing over his notion of a "miscarried" experiment. Suppose Dee is permitted to perform her tasks in appropriately controlled circumstances without distractions, and assume it is true that she will now be able to correct calculations *unless* unforeseen, error-inducing influences interfere. (We thus assume that the experiment has no mistakes in design or implementation.) Surprisingly, Dee makes completely elementary errors in calculating. I think Wittgenstein would say this experiment has "miscarried" despite the experimenter's care and at least general cooperation from the environment. For the question comes up whether Dee is an appropriate subject for the experiment, since it is doubtful that such elementary errors could be made by a "normal" adult with average educational background. Yet the fact remains that no *experimental error* has occurred. (Investigation might show that

Dee is suffering from exhaustion without realizing it, or her errors might be "inexplicable" (III, 68). Without an explanation, though, our experimenter would do well to find another subject.)

Consider an analogous case where Dee is a mathematical genius and our aim is to see how *quickly* she can calculate (correctly). As before, she makes many errors, perhaps including "sophisticated errors." Therefore, the experiment *miscarries*, not because it is improperly designed or implemented, but because a *background* assumption is not satisfied, namely, that under the new specified conditions a cooperative genius can and will perform many calculations without error. (Only our being able to count on that would have made it possible to find out the speed at which Dee can calculate.) Though there is appropriate experimental design, though implementation secures the controls laid out in the design, and though the subject has presumably been suitably chosen, interference remains possible, if improbable. So to say an experiment "miscarries," unlike saying it involves "mistakes," is not negative criticism. A miscarriage occurs when an attempt at experimentation "misfires": an intended experiment begins, but cannot be brought to completion until the failure of background assumptions is remedied.

IV. AN ATTACK ON APRIORISM

As we noted, Wittgenstein rejects any mentalist conception of our relation to a proof which we "grasp" or "understand." He holds that proofs are "surveyable" and that our ability to understand them rests, at least, on this fact. According to remarks in Section III, then, our grasp of proof derives from logical features which distinguish proofs from experiments, errors in an alleged proof from errors in an experiment, and so on. We are now ready to examine Wittgenstein's apriorism by focusing on two features that "surveyability" entails, features that are clearly fundamental for him. These are the holism of proofs and their unconditionedness in contrast with experiments.

Now the feature of being "laid clear to view" – of having perspicuous parts and overall organization – is different from holism and from unconditionedness. A clear proof is reproducible "as a whole" in that it can be copied, if carefully, without error. But this does not imply that copying a proof entails copying its *result*. That is, perspicuity does not imply holism. On the other hand, holism does not guarantee perspicuity. The lack of it might make it impossible for us to reproduce a proof, but does not entail that any proof is not holistic. Put otherwise, from the fact that (copying) any proof must include (copying) its result, nothing follows about whether or not that proof is laid clear to view.

If, as note 10 suggests, proofs' unconditionedness is deducible from their holism, then the preceding remarks imply that unconditionedness is no guarantee of perspicuity. Thus it is generally possible that there be unconditioned kinds of intellectual performance, that is, performance independent of particular empirical conditions, whose details or organization are less than wholly clear. And why not!? Significantly, an unconditioned performance, for example construction of a proof, is in a reasonable sense "free of empirical elements," since no *specific* empirical claims must be true in order for that performance to occur. Even the most unperspicuous of constructions, then, can in principle satisfy what I have said are fundamental requirements in Wittgenstein's conception of *bona fide proof*-construction. (How much unclarity a genuine proof can, as it were, tolerate is not a question Wittgenstein addresses.)

Given these preliminaries, let us turn to the topic of apriorism about proof-content. We have seen that such a thesis would follow from Wittgenstein's demand that mathematical truths and all other "proof-contents" implicate no empirical elements (Section II). Yet, we saw, he may want to allow for at least some proofs whose details are not wholly clear. That Wittgenstein might, by the above reasoning, allow for unperspicuous *a priori* proofs supports those who question the standard claim that an unperspicuous construction like the proof of the Four Color Theorem must thereby contain empirical content. The claim has been used, for example by Thomas Tymoczko, to conclude that permitting the Appel–Haken–Koch construction to count as "proof" alters our very concept of proof.[12] Against this, Detlefsen and Luker have argued that the unsurveyability, that is, unperspicuity, of a construction is not "at bottom responsible for its being based on empirical considerations," if it is.[13]

This much of the substance of what Detlefsen and Luker (1980) have to say is consonant with the interpretation I have presented of Wittgenstein's views. But the Detlefsen–Luker position derives from an argument to show that, assuming a certain conception of proof, *all* proofs contain empirical content, whereas only some are unsurveyable. (Unsurveyability, that is, unperspicuity, will therefore not explain why a proof includes empirical elements, if it does.) Detlefsen and Luker argue that, *if* proofs are conceived as being, in their phrase, "self-sufficient," then every proof must incorporate empirical content. This argument is not so easily overturned, since, for one thing, the idea of self-sufficient proof they develop can very naturally be associated with apriorist accounts like, for example, that of Descartes. In addition, Wittgenstein pretty clearly accepts that idea himself.

What is "self-sufficiency"? It is the property any reasoning has if it is "a unit of reasoning that *needs nothing outside of itself to carry conviction*";[14] in a sense, a "self-sufficient epistemological packet."[15] "Self-sufficiency," then, implies that, if a proposition must be believed, or some reasoning endorsed, by a potential prover in order that she/he gain "*confidence or conviction*" about a proof's result, then that proposition, or that reasoning, is *part* of the relevant proof.[16] One could argue that a prover must always believe, at least implicitly, that *these* signs which she/he uses are not merely hallucinated, and hence that "self-sufficiency" entails that propositions about individual provers and the signs they use are part of any proof's content. Of course this is implausible. But further assume that your and my constructions embody or constitute the same proof only if their propositional content is the same. Then the evident facts about human proof-construction land us with the wild conclusion that constructing the "same proof" depends on our making it together using identical signs!

Obviously, Detlefsen and Luker want to avoid such strange conclusions, and their intention is to appeal to a reasonable notion of self-sufficiency which can pose a real threat to apriorism. Indeed, they are careful in specifying two types of empirical claims which, they believe, are required for conviction about even the simplest of proofs, for example, Gauss's proof that the sum of the first 100 integers is 5,050.[17] The first sort of required empirical believe, illustrated for the case of calculation, is "(c) that the computing agent correctly executes the program."[18] The second sort of belief asserts "(d) that the reported result was actually obtained."[19]

To become convinced of Gauss's proof, I must believe that the relevant proof-construction is obtained by correct employment of the "program" used to guide its making. I must belief the construction results from correctly *adding* fifty pairs of integers to obtain "101" in each case and correctly *multiplying* 101 by 50 to obtain

"5,050." If I did not believe that the correct sum is 101 in every column and believe that the correct product of these sums is 5,050, I would not ordinarily become convinced that Gauss's statement has been *proved*. I must believe that the system of procedures for addition and multiplication has been properly "executed." The second type of empirical belief I need in the present example is that Gauss's result, "5,050," is the result obtained in the construction in question, that it is the result of the activity of constructing in which I have engaged.

Detlefsen and Luker are suggesting that a (rational) calculator cannot become convinced that a construction C establishes result R unless she/he believes both that no arithmetical errors occur in C and that, in the pertinent language, the result obtained or found in C is "R". If I worry that my fiftieth column – the one where we add "50" and "51" – contains an erroneous "sum," I then have to doubt whether the construction is proof of Gauss's claim. If I am uncertain whether "5,050" resulted from multiplying "101" by the number of columns, that is, uncertain about what result was found, I will be uncertain whether the construction proves *any* result. These claims appear to be reasonable. Accordingly, Detlefsen and Luker seem right to say that the belief that a specific construction is a proof – *if* this species of belief is possible – rests on empirical beliefs. In that event, they also would be right that self-sufficient proofs – *if* proofs be specific constructions – must have empirical content.

My qualifiers have a point. I think Wittgenstein does not view the conviction that X literally *is* a proof as strictly identifiable with a belief about specific constructions. I think he parts ways with Detlefsen and Luker, not by denying that proofs are epistemically self-sufficient, but by construing "belief in a proof" as belief about the demonstrative status of a *type* of construction, and this naturally suggests the view that proofs themselves are construction-types rather than spatial or temporal configurations which humans literally *make*. In Section V, then, we probe more deeply the notion of "mathematical conviction." We also present exegetical support to show that the content of our demonstrative knowledge precludes, and is independent of, information about provers, configurations of sign-tokens, and other empirical facts.

V. APRIORISM: WITTGENSTEIN'S DEFENSE

Wittgenstein says, "We do not accept the result of a proof because it results once, or because it often results. But we see in the proof the reason for saying that this *must* be the result" (III, 39). Recall his interlocuter (Section II) who asks "Must this *always* be the result, in that case?" and to whom Wittgenstein responds, "Of course" (III, 35). Learning what the result always is or must be is learned by grasping the proof, which provides "the reason" that R, say, is the result. Is learning that R is the (correct) result of a proof even partly a matter of learning that "R" is the result of a specific construction C? To answer this, note that in the first passage cited above, Wittgenstein maintains that the fact that "R" results in specific constructions is not our evidence that R is the result. Rather, he says, the "reason for saying" that R is the result may be seen "in the proof." But, we must inquire, what is "the proof"?

As proposed at the end of Section IV, "proof" refers to construction-types: "The proof (the pattern of the proof) shews us the result of a procedure (the construction); and we are convinced that a procedure regulated in *this* way always leads to this configuration" (III, 22). Proof convinces us that such-and-such a pattern (type) of procedure, "a procedure regulated in *this* [kind of] way," must lead to "this [kind of] configuration." "This configuration" cannot refer to a *token*-construction, since no *one* construction "always" results upon applying a procedure. In any event, III, 22, helps us understand why Wittgenstein says that a proof "must originally be a kind of experiment – but is then taken simply as a picture" (III, 23) and that a proof is "the picture of an experiment as it practically always turns out" (VII, 18). The latter remarks and III, 22, are mutually supportive if we observe that taking a proof as a "picture" is just taking it as the "pattern" shared by most of our experimental attempts to embody the proof in concrete constructions.

If proofs are types of construction, then, unlike particular states of affairs, "We construct the proof once for all" (III, 22). And constructing it "once for all" enables a proof to play its role as "model" or "rule":

> The proof is our model for a particular *result's being yielded,* which serves as an object of comparison (yardstick) for real changes.
>
> (III, 24)

I learned empirically that this came out this time, that it usually does come out; but does the proposition of mathematics say that? I learned empirically that this is the road I travelled. But is *that* the mathematical proposition? - ... What relation has it to these empirical propositions? The mathematical proposition has the dignity of a rule.

(I, 165)

Hence, grasping a proof (i.e. proof-type, pattern, picture) depends on grasping something general – not how "this came out this time" (I, 165), but "what OUGHT to come out" (III, 55).

Support for my account is present in numerous passages: for example, "A proof must be our model, our picture, of how these operations have *a result*" (III, 24) and "the conviction produced by a proof cannot simply arise from the proof-construction" (III, 35), that is, the *token*-construction. Moreover, Wittgenstein contends that a proof is an available model for the proper *reproduction* of potentially many of its own "proof"-tokens: "Now if a proof is a model, then the point must be what is to count as a correct reproduction of the proof" (III, 44). In this passage "reproduction of the proof" refers to its repeatable "embodiment" in token-constructions (cf. III, 69).

To digress briefly, I mentioned (note 8) that Wittgenstein emphasizes the easy reproducibility of proof, often by talking about "perspicuity." Instead of indicating a commitment to an epistemic notion of surveyability, I would maintain that the emphasis on clarity in constructions has as its purpose reference to the one–many relationship I have argued for between proofs and their tokens. Admittedly, however, a great deal more exegetical work needs to be done to settle this issue. It is also crucial at this point to see that identifying proofs as types does not commit Wittgenstein to the platonistic view that proofs are eternal, abstract entities. Evidently, his claim that we "construct" proofs "once for all" (III, 22 above) would be literally false were platonism true, in contrast with the claim that we "discover" proofs once and for all. Yet, as is known, we frequently find Wittgenstein suggesting that talk about mathematical "discoveries" can be misleading if not properly interpreted, but never, to my knowledge, questioning our talk about the "construction" of proofs.

I favor the view that understanding that R is the result of a proof is not a "matter" of, that is, does not include, discovering that "R" is obtained in some token-construction. A proof, we explained, is a type or pattern which comes to regulate procedural or constructive activities. Since proofs are also rules governing the application of concepts, they also guide judgments about the meaningfulness of empirical descriptions. This general, regulative nature of proofs shows, according to Wittgenstein, that mathematical conviction or understanding must pertain to content which is radically different from that of empirical descriptions. Because mathematical content is of a distinct sort from empirical, mathematical propositions on Wittgenstein's account essentially include their own potential applications: "what interests us" about what a proof "convinces us of" is "not the mental state of conviction, but the applications attaching to this conviction" (III, 25). Wittgenstein refers to consequences for action which accepting a mathematical result will have as the result "serves as an object of comparison (yardstick) for real changes" (III, 24).

Hence, whether such and such a sequence of operations has led us to conclude "R" is in no way *part* of mathematics. We learn the mathematical result when we grasp, what *otherwise* would be an *experiment*-with-signs, a *standard* or *rule*. Wittgenstein's general philosophy, then, implies that learning the mathematical rule is learning a standard for *sameness or difference*, in this case the sameness or difference of various constructive activities. To learn this is not to make a "judgment" in the empirical sense, but to accept a "*foundation*" (VI, 46), a "procedure on which we build all judging" (VI, 28).

These remarks have supported the claim that understanding that X is a proof does not *include* any beliefs or convictions about token-constructions. This does not establish the contention that grasping a proof is *independent* in a significant way of beliefs or knowledge[20] about token-constructions, a thesis we now go on to argue for.

Let us first return to the idea of the conceptual creativity of proofs mentioned in Section II. Proofs create new concepts on Wittgenstein's view. The proof that 200 apples and 200 apples yield 400 apples, "defines a new concept: 'the counting of 200 and 200 objects together'.... The proof *defines* 'correctly counting together'" (III, 24). We are given "a *new* paradigm for determining when objects have been counted correctly (III, 41).

Can I say: "a proof induces us to make a certain decision,

namely that of accepting a particular concept-formation"?
Do not look at the proof as a procedure that *compels* you, but as one that *guides* you. – And what it guides is your *conception* of a (particular) situation.

(IV, 30)

In the present example, it guides our conception of what the result *must* be if one *correctly* counts 200 apples, and then 200 more: "The concept [of the sum of 200 and 200 objects] is altered so that this *had* to be the result" (IV, 47).

Wittgenstein also describes concept formation in terms of newly established connections: "I should like to say: the proof shews me a new connexion, and hence it also gives me a new concept" (V, 45).[21] Earlier in the same set of texts, he refers to a case in which a newly created connection seems plausible: "An equation links two concepts; so that I can now pass from one to the other" (V, 42). For instance, the equation "200 + 200 = 400" expresses the connection of identity holding between the sum of 200 and 200, on the one hand, and 400 on the other. This connection also holds between our *concept* of the sum of 200 and 200 and our *concept* of 400, but not prior to the proof. New individual concepts likewise emerge in that our concept of the sum of 200 and 200 now incorporates the idea of that sum's identity to 400, and similarly the concept of 400 includes 400's being identical to the sum of 200 and 200.

Now Wittgenstein considers an objection to his claim that mathematical proofs are creative, and his reply supports the idea that grasping proof rests on, if anything, human *action* rather than "inner experience." The objection is: "'all this is involved in the concept itself,'" that is, is involved *prior* to the proof, so the proof is not truly originative of concepts – what the proof reveals was already "there" in the concepts operant in the proof (cf. "Mathematics is analytic"). Wittgenstein's reply is that "what that [namely, "all this is involved in the concept itself"] means is that we incline to *these* determinations of the concept. For what have we in our heads, which contains all these determinations?" (VII, 42).

Our concept of the sum of 200 and 200 alters only when its necessary connection with the concept of 400 has been proved. And our inclination to accept the demonstration, and therein the new concepts, is not explicable in terms of prior conceptualization or psychological processes (cf. "in our heads"). Wittgenstein is insisting on the fact of concept formation via proof and its independence of the mental (recall App. II, 1–8, and VII, 18). Beyond rejecting mentalist explanations of *a priori* knowledge, he further emphasizes that our acceptance of proofs is not in *itself* an "experience." The sole criterion of our accepting a proof lies in our action, in what we are inclined to *do*. Thus:

> "To give a new concept" can only mean to introduce a new employment of a concept, a new practice.
>
> (VII, 70; cf. VI, 38)

> (1) The word that is written there (or ... was just now pronounced or etc.) has 7 sounds. (2) The sound-pattern "Daedalus" has 7 sounds.
> The second proposition is timeless.
> ...
> The timelessness of the second proposition is not e.g. a result of the counting, but of the decision to employ the result of counting in a particular way.
>
> (VI, 36)

That we grasp a proof is reflected in a new employment of an old concept. The concept of the sound-pattern "Daedalus" will now be used only for patterns containing just seven sound-elements. Patterns not like this *cannot*, we will say, be the same as "Daedalus," however much similarity is present. It is what we do with a concept that reveals what concept it is.

Wittgenstein argues for the same point in his references to how proofs teach us "internal relations":

> For the internal relation is the operation producing one structure from another, seen as equivalent to the picture of the transition itself – so that now the transition according to this series of configurations is *eo ipso* a transition according to those rules for operating.
>
> (VII, 72)

During a proof we come to grasp an internal relation by being persuaded to "idealize" some physical transformations which signs underwent in our token-constructions. That is, we come to *see* what otherwise would just be a causally related stream of transitions *as* a "picture of the [correct] transition itself." But this "change in aspect" does not arise in lawlike fashion from what we may happen to experience while studying a construction. Nor is the new "aspect," the "picture of the transition itself," reducible to something in token-constructions of which we have "mental experience" (App. II, 2).

What, then, is the criterion for new concepts' having been acquired, or for our understanding of proofs, if it is not an experience with which specific constructions have provided us? Wittgenstein claims that the applicable criterion mainly concerns *future* action. Now we will take it that a token-configuration can *be*, that is, be *said* to be, *correct* only if we judge it to instantiate the demonstrated type, the "picture" we accept (cf. I, 61-3).

To put it differently, Wittgenstein says that mathematics yields certain "prophecies." The prophecy a mathematically proven proposition yields "does *not* run, that a man will get *this* result when he follows this rule in making a transformation – but that he will get this result, when we *say* that he is following the rule" (III, 66). The mathematical rule is our "picture" of the correct transition, a type of "memorable configuration," yet its very content can be understood only by reference to what we do:

> the question is: What do we call a "memorable configuration"? What is the criterion for its being impressed on our minds? Or is the answer to that: "That we use it as a paradigm of identity!"?
>
> (III, 9)

The correctness of a type of construction has no explanation. At best, we may observe that we use proofs and their results as we do: "I have a definite concept of the rule. I know what I have to do in any particular case.... I say 'Of course'. I can give no reason" (VI, 24; cf. 1, 63).

To summarize, Wittgenstein holds that understanding a proof in a sense "transcends" our grasp of specific constructions, for our conviction that such and such is a way of proving X is not identifiable with, and does not involve, belief about spatial or temporal constructions. Wittgenstein also holds the stronger view that knowledge of proof is *independent* to a significant degree of beliefs about particular configurations. Still, it is not as if one could learn a proof with no appeal *whatever* to the latter. In addition, some of the above citations may themselves seem to intimate that "grasping a proof" at least indirectly includes a reference to our knowledge of tokens. We have seen, after all, that grasping a proof is accepting a "picture" *of* a particular type of configuration as representing "the transition itself," that is, as representing the correct transition. One may with some plausibility take this to mean that, as a matter of psychological fact, a *specific* construction of the requisite type must serve as the object of "picturing" activity.

Now whether the construction that serves this purpose is perceived in space or imaginatively visualized does not, of course, affect the point. But this line of thought seems to me to over-emphasize Wittgenstein's picture-analogy. There is no textual evidence supporting the idea that his "proofs" must be, as it were, "still shots" of visually perceivable configurations. Many of his examples are visual, yet he does not ever claim that proof must depend upon visualization. (If it must, then, arguably, persons without sight cannot understand proofs! It is hard to believe Wittgenstein would make what is presumably a factual error, or risk this kind of error, in the name of analyzing concepts!) The interpretation at issue would be natural if one thought Wittgenstein *required* proofs to be perspicuous (and took perspicuity as an essentially visual notion, as I do). Apart from the factual difficulty observed, there is the added difficulty of sustaining the claim that Wittgensteinian proofs must be perspicuous. I have argued otherwise (e.g. in Section III and note 8), and tried to show the usefulness of Shanker's idea that logical features of proof, features related to the logical distinction between proofs and experiments, are central in Wittgenstein's philosophy of mathematics.

I hope the preceding sections of this essay have helped to show the fruitfulness of Shanker's idea. That idea may eventually force us to modify some of our favored interpretations of Wittgenstein on mathematics and logic. Certainly, I think the time has come to de-emphasize the visual elements in his account. Again, however, that is not to say that our knowledge of proofs is not dependent on contingent facts about linguistic practice and about our "natural history." Surely it is. And nothing I have said here precludes such knowledge from being dependent upon, for instance, the fact that at least *some* token-constructions are perceptually and imaginatively available. Likewise, what I have said here has left open the possibility of reconciling what seems to be Wittgenstein's "dual" view of proof, namely, proof as standard (the "hardness of the logical 'must'") and proof as a variable, practical phenomenon.

NOTES

1 See Wittgenstein (1956). Unless otherwise noted, we use section and paragraph numbers in the text to refer to passages in this work.
2 See Kreisel (1958), pp. 136 and 142 respectively.
3 See Wittgenstein (1958: 230).
4 We need the phrase, "in principle at least," in order to allow for cases where mathematics plays essential role in giving the content of an experiment. The point is that members of the genus *experiment* allow generally of experiential description.
5 Passages denying that mathematical statements have any empirical content are not unusual in Wittgenstein (1956), or for that matter in Wittgenstein (1976). And Wittgenstein is quite often intent on discrediting the simplistic empiricist who says these statements express "'anthropological' propositions saying how we men infer and calculate ..." (1956: III, 65). Note that, throughout, points about mathematics and mathematical truth apply equally to logic and logical truth. In the citations, some examples are mathematical, some from logic.
6 See Wittgenstein (1956: III, 37), where Wittgenstein distinguishes between the *employment* of, and the *determination* of, a sense of meaning. In describing experiments we *employ* words having meaning, while in describing what the content is of a demonstration or proof we must convey a meaning newly *determined.*
7 See Shanker (1987).
8 It sometimes appears that *übersehbar* ("surveyable") and *übersichtlich* ("perspicuous") are used interchangeably in Wittgenstein (1956) (e.g. III, 11, 42–3). But the epistemic requirement that proofs be perspicuous or "*plain to view*" (III, 42) is not as central to Wittgenstein's development of the proof/experiment distinction as is his network of logical points of contrast between proofs and experiments. It is not even clear that perspicuity need be taken as a Wittgensteinian constraint on proof.

We might hold, for example, in the context of studying his critique of logicism (cf. 156: III), that he means to contrast a set of perspicuous proofs, namely, ordinary calculations to Russell's set of unperspicuous ones. We need not deny here that in many contexts proofs whose details are presented "plain to view" possess significant advantages, advantages of a sort emphasized in Wittgenstein's critique of Russell. Thus, perhaps in order to construct any proofs there must first be some perspicuous proofs enabling us to establish comprehensible and orderly demonstrative procedures. From this it does not follow that the concept *proof* includes the property of being perspicuous.

9 See Shanker (1987) e.g. pp. 128ff.
10 There seems to be a connection between the conditionedness of experiments and the fact that they are holistic. To put it differently, the holistic nature of reconstructions of proofs seems responsible for the fact that they are not conditioned in the way experiments must be. I suspect, but cannot prove here, that the unconditionedness, in the present sense, of proofs is deducible from their holism.

11 See note 4.
12 See Tymoczko (1979: 59–60, 70ff).
13 See Detlefsen and Luker (1980: 814).
14 ibid., p. 810.
15 ibid., p. 811, fn 7.
16 ibid., p. 810.
17 ibid., pp. 807–8.
18 ibid., p. 808.
19 ibid.
20 Wittgenstein argues in numerous passages that the distinction between (mere) belief and knowledge cannot coherently be made in the mathematical case. As I noted in Section III, he does not allow for one's having just a partial understanding of a proof. He maintains that, in this event, one has not as yet grasped the whole proof and, in a sense, cannot yet have formed *beliefs* about such and such a proof. For belief requires an object, and in the case in question one has no clear view of the proof-object. Once one has such a view, however, one has already seen that the relevant construction "embodies" proof, and thus one has knowledge of a proof. Again, see Wittgenstein (1956: App. II, 1–8, VII, 18 and 26) and see, for example, Lectures XIII–XIV in Wittgenstein (1976).
21 The point of "hence" here remains somewhat obscure to me. The passage, however we read it, is not obviously consistent with others. The ensuing comments, however, should be clear enough for us to proceed.

REFERENCES

Detlefsen, M. and Luker, M. (1980) "The Four-Color Theorem and mathematical proof," *Journal of Philosophy* 77: 803–20.

Kreisel, G. (1958) "Wittgenstein's remarks on the foundations of mathematics," *British Journal for the Philosophy of Science* 9: 135–58.

Shanker, S. (1987) *Wittgenstein and the Turning-Point in the Philosophy of Mathematics*, Buffalo, NY: State University of New York Press.

Tymoczko, T. (1979) "The Four-Color Problem and its philosophical significance," *Journal of Philosophy* 76: 57–83.

Wittgenstein, L. (1956) *Remarks on the Foundations of Mathematics*, ed. G.H. von Wright, R. Rhees and G.E.M. Anscombe, trans. G.E.M. Anscombe, Boston, MA: MIT Press.

—— (1958) *Philosophical Investigations*, ed. G.E.M. Anscombe and R. Rhees, trans. G.E.M. Anscombe, 2nd edn, Oxford: Basil Blackwell.

—— (1976) *Lectures on the Foundations of Mathematics, 1939*, ed. C. Diamond, Ithaca, NY: Cornell University Press.

5

ON THE CONCEPT OF PROOF IN ELEMENTARY GEOMETRY

Pirmin Stekeler-Weithofer

SUMMARY

The concept of proof used in elementary synthetic geometry relies heavily on the fact that we really can fulfill certain intentions of forming solid bodies. Although we say that a body is solid if it does not change its form by changing position, there is no concept of form independent from that of solidity. The conditions of satisfaction of the intentions of forming solid bodies must be formulated in a holistic way using a *descriptive* language. We must control a plan to construct bodies or pictures of certain forms by *real* observations. An analysis of the basic concepts of idealization and abstraction shows, then, that any justification of Euclidean geometry or general kinetics rests on experience. It cannot be *a priori* in an absolute sense, as Kant, Dingler, and the German constructivists seem to claim.

I. ELEMENTARY SYNTHETIC VERSUS AXIOMATIZED AND ANALYTIC GEOMETRY

What I shall call "elementary synthetic (plane) geometry" is essentially the practice of elementary school geometry. There we learned to draw two-dimensional figures on a sheet of paper, using certain fixed means such as a sufficiently plane surface to draw upon, a straight rule with two marked points on it (representing the chosen unit length), and a circle. And we "proved" geometrical statements by describing geometrical constructions represented by such drawings, together with some comments on possible movements of surfaces or on possible mirror-mappings. It is this special concept of "proof" that I propose to call "demonstration." In this way we prove, as it well known, theorems like those of Thales or Pythagoras in elementary synthetic geometry. Hence, elementary synthetic geometry is not an axiomatized system – despite the fact that Hilbert's famous book *Foundations of Geometry* tried to reconstruct an axiomatic theory of geometry just as Peano had done for arithmetic, and Zermelo and others for set theory. In a fully axiomatized system a "proof" is just a series of deductions, that is, a series of applications of the rules of *first*-order predicate calculus, where one starts with a set of formulas called "axioms." Hilbert's geometry is no such theory for the following reasons: neither his Archimedean axiom nor his versions of the completeness axiom are formulated in a first-order language; and his proofs are not all first-order deductions, but are often *semantic* proofs.

Euclid's "axioms," on the other hand, are *general* (definitional) meaning postulates for any reasonable talk about an (abstract) "equality" and/or "ordering," that is, they give the defining conditions of an equivalence relation or a linear ordering. Hence, in any particular case it must be postulated or shown that these axioms are fulfilled, for example, for the congruence relation, for the relation of similarity, or for the definitions of "equality" of proportions. Axioms are not at all intended to define geometrical entities and truth conditions "implicitly," neither are Euclid's "postulates" to be viewed as axiomatic premises of formal deductions. They are all, including the famous postulate of the uniqueness of parallels, basic assumptions about geometrical constructibilities, general presuppositions of the concept of demonstration in synthetic geometry.

In *analytic* geometry we *represent* what we call geometrical "points," "lines," "surfaces," and "volumes" as sets of pairs or triples of (real) numbers, where real numbers are defined as converging sequences of rational numbers, or, if one prefers, as Dedekind-cuts in the rationals. Such a system of higher arithmetic" is a *model* for geometry proper, in the sense that we can *replace* genuine geometrical demonstrations of elementary synthetic geometry by formal proofs based on logical (set-theoretic) and arithmetical definitions. If we want to understand the meaning of such proofs, we should always remember *why* certain logical and arithmetical operations on sentences (containing free or bound variables or names of (sets of) numbers) *really* can be interpreted as geometrical proofs.

Whereas *elementary* (plane) geometry only talks about *straight lines* and *circles*, in generalized analytic geometry all kinds of curves, surfaces, and volumes can be represented by certain "*point-sets*" or *functions*, and the calculus of differentiation and integration can be used. It is understandable that in mathematics we almost always use analytic means to talk about geometrical entities. But we may ask: why does it make sense at all to call purely arithmetical and set-theoretic entities in the "spaces" R^3 or R^2 "points," "lines" or "curves," or "surfaces" or "volumes"? Only if we distinguish between the three systems of geometry, namely elementary synthetic geometry, systems of axiomatized geometry, and analytic geometry, with their different concepts of truth and proof, can we analyze the relations between them. Indeed, elementary synthetic geometry provides us with the projection rules to translate the "formal" and "ideal" ways of talking in mathematics of spaces and geometrical forms into the "external" concepts of real physical space and the (geometrical) forms of real physical bodies.

II. THE PYTHAGOREAN PROGRAM AND EUDOXUS' PROPORTION THEORY

A brief look at the history of mathematics shows that geometry is not just analytic geometry. On the contrary, one of the most important historical steps in mathematics consisted in the formulation of the *program of arithmetization of geometry* as it was put forward by the early Greek mathematicians, the Pythagoreans. Their basic goal, indeed, was to represent sentences and demonstrations of synthetic geometry by sentences and proofs in arithmetic. The Pythagorean Theorem, for example, shows how we can characterize a rectangular triangle by the equation $x^2 + y^2 = z^2$. Yet, the famous proofs of the irrationality of the proportion $a:b$ of the lengths of a side a and a diagonal b in a square or a pentagon made it difficult to achieve the original Pythagorean goal, since the first theory of proportion seemed to be a theory of *ratios $n:m$ between integers*. It is quite probable that these "irrationalities" were not detected before the development of a more general theory of proportion, defined by the so-called "anthyphairesis" or "antanairesis":[1] If A, B are arbitrary sizes, especially lengths, and if $A > B$, then $A:B$ is defined by the sequence of numbers $[n_1, n_2, n_3, \ldots]$ such that n_1 is the maximal number with $n_1 \cdot B < A$ (i.e. B goes n_1 times into A, but not $n_1 + 1$ times), n_2 is the maximal number with $n_2 \cdot (A - n_1 \cdot B) < B$, the new remainder goes n_3 times into the old one, and so on.[2] This procedure is just Euclid's algorithm if $A:B$ is a ratio $n:m$; it does not stop if and only if the proportion is not rational.

PROOF AND KNOWLEDGE IN MATHEMATICS

Probably because of the considerable technical difficulties of such a theory (where a proportion is "named" or "represented" by a series of natural numbers) Eudoxus developed a beautiful new theory, which is reported in Euclid's Book V. His definition of a proportion is extremely general and shows a perfect understanding of the logical constitution of "new" abstract entities. He begins by laying down conditions for an arbitrary domain of "sizes," that is, a domain of (abstract) entities, for which at least multiplication by integers and a *linear Archimedean ordering* is defined; that is, if A, B are sizes, then $A < B$ or $B < A$ or $A = B$ must be true, and there always must be an integer n such that $B < n \cdot A$. (Cf. Euclid V, Def. 1–4.) Examples of such domains are the integers, the lengths of straight lines, and the sizes of surfaces or volumes. If A, B are sizes of some domain of this sort, then the expression "$A:B$" is "semantically well-formed." If C, D are sizes of the same or different domains, then Eudoxus and Euclid define equality $A:B = C:D$ and inequality $A:B < C:D$ by the following formal (definitional!) truth-condition: If there are integers n, m such that $m \cdot A < n \cdot B$ but not $m \cdot D < n \cdot C$, then $A:B < C:D$ holds. If neither $A:B < C:D$ nor $C:D < A:B$ holds, then $A:B = C:D$ holds. (Cf. Euclid V, Def. 7 and 5.) To say that a semantically well-formed expression "$A:B$" "names" a proportion just means that truth-conditions for equalities between such expressions are defined. Together with the identity $n \cdot (A:B) = (n \cdot A):B$, which defines multiplication of proportions by integers, Euclid (in Book V) can easily prove that on the basis of this definition the proportions of one domain of sizes (or even a set of such domains!) themselves form a domain of sizes in the above sense.

This theory of proportion is used, then, to prove the important geometrical statement that the proportions $a:b$ of the ground sides of triangles (or parallelograms) with fixed height are the same as the proportions $A:B$ of the surfaces. The easy proof can be found in Euclid Book VI, 1. From this fact we almost immediately arrive at some very important geometrical propositions such as the famous theorem of Desargues, telling us the basic features of central projection of geometrical figures, that is, of geometrical similarity. (Cf. Euclid VI, 2ff.)

This classical way into a theory of geometrical similarity shows quite clearly the externally (geometrically, pragmatically) the following stability conditions for "real" (triangular) plane surfaces and straight lines on solid bodies are *presupposed*: we can *compare* the sizes of the surfaces and lines and *add* the surfaces and lines. This is done, of course, by moving (good) *copies* of the surfaces (or of the lines) on solid bodies and fitting them one to another. We can also fit them into one or more figures like we do in a Tangram game: if the same pieces make up a surface A and a surface B (a line a and line b respectively), we say that A and B (a and b) are of the same size. The criteria used to judge a good, solid copy are that good copies should not *grow* or *shrink* when moved. This we check in *two* ways. First, we compare (good) copies of the same original, they should always fit into the same (good) "hollow form," that is, they should remain "congruent." Second, we assume that if we have two (let us say triangular) plane surfaces A and B (or two straight edges a, b), then there is a number n such that we can join n copies of A (or a) to form a bigger surface than B (a longer line or edge than b). The latter condition does not follow from the first: one can imagine a situation where all "copies" of a small surface A "are congruent," but where we cannot outgrow B by copies of A since they "shrink uniformly" if we move them on B away from some fixed center. In such a case we could view B as the "space" in which the copies of A are moved such that they "shrink" and yet have a "constant form" (relative to each other and their hollow forms). In other words, the condition that sizes form an *Archimedean* linear ordering is a very important pragmatic condition for a *relative* judgment of what it means for a body, surface, or edge to be "solid," to have a stable form. We judge this in a holistic way, namely by dividing "all" bodies and their surfaces and edges into two classes: the (more or less) solid ones and the ones which change their form in time or when moved. Change needs something to measure it as change. This is a truism. This "something" is in our case a whole class of things and their copies, not, as many people seem to think, some special body as the meter in Paris or some nonthing as "the real empty space."

III. FURTHER DEVELOPMENTS IN THE ARITHMETIZATION OF GEOMETRY

Despite their theories of proportion the Greeks altogether lacked a purely arithmetical definition of even the number field of rational numbers, not to speak of the reals. It is just a big mistake to claim that Eudoxus' proportions were equivalent to Dedekind-cuts.

There is not even a general definition of addition and multiplication of ratios $n:m$ of integers n, m to be found in Euclid, not to speak of the addition and multiplication of general proportions. What we do find is the addition and subtraction of lengths and an operation of multiplication $a \cdot b$ on lengths, which was known as the inverse operation of "division" $a:b$. Moreover, we obviously can divide proportions further, that is, we can calculate with expressions of the form "$(A:B):(C:D)$" (and so on). Euclid knew only *some important* but (in comparison with the rules of the real number field) rather restricted formal calculations with expressions of the form $A:B$. What we find are rules such as the following: $(A+/-B):B = (C+/-D):D$ holds iff $A:B = C:D$ (V, 17, 18); $n \cdot A : n \cdot B = A:B$ (V, 15); and the important theorem (V, 22) for multiplication, if $A:B = D:E$ and $B:C = E:F$, then $A:C = D:F$.

Eudoxus and Euclid did not seem to see the possibility, however, of *identifying* a proportion $a:b$ of lengths with a length c, if c is the solution of the equation $a:b = c:e$ and e is a fixed unit length. This is astonishing, since they knew that there is a geometrical construction of the third proportional x, solving the equation $a:b = x:d$, if a, b, c are given lengths. This identification would have almost immediately shown that and how the multiplication of lengths $a \cdot b$ leads to a length, not (only) a size of a surface, as a result. So we would have arrived at the space of Pythagorean or Euclidean lengths, which algebraically form a *field*, if we only distinguished positive and negative lengths. Perhaps just because of its formal rigor Greek mathematics did not take this step.

A much later development in the arithmetization of geometry was Descartes' simple but nevertheless powerful idea of representing points on a plane or in space by pairs or triples of lengths, which were the results of their orthogonal projection onto a fixed coordinate system. This idea immediately shows how calculations with positive and negative lengths can be interpreted geometrically. Note that I still speak of *lengths*, not *numbers*, here, since a construction of Pythagorean or Euclidean "numbers" rests either on the method of constituting a field as an *algebraic extension* of the rationals by a description of a system of equations, or on the definition of the real number field in the way given by Dedekind. In the latter case we can define the *subfield* of the Pythagorean or Euclidean real numbers by some *differentia specifica*. Descartes did not have a concept of the real numbers, nor did he know the method of constituting domains of entities as "fictitious" solutions of certain equations. Not even Leibniz or Newton could really *define* infinitesimals and reals, but thinking geometrically, on the basis of "intuition," they "hoped" that the domains they needed for their calculi could be assumed. Well, the hope was not unfounded since there were geometrical "equivalents" or at least "pictures" for them. Nevertheless, in comparison with Eudoxus' definition of proportions their concepts of infinitesimals were rather vague. But it would be entirely wrong to blame Leibniz or Newton for this. On the contrary: the extreme logical precision which did not allow "identifications" between proportions and lengths made Eudoxus' definition less powerful than it could have been.

IV. LOGICAL ANALYSIS OF ABSTRACT ENTITIES

In the historical development of mathematics the "belief" or "idea" that certain domains of abstract things and properties "exist" very often *precedes* a logical clarification of the (linguistic) *constitution* of such domains of mathematical discourse. This might be the reason why mathematical ways of speaking have produced many uncritical (conventional, naive) "beliefs" in Pythagoristic ontologies, in domains of "real forms" lying behind or below geometrical experience, in some "real infinity" and other properties of "real" space and "real" time, or, to take other examples, in some "real" domain of sets or even of so-called "possible worlds." If not viewed as rather general (hence necessarily vague and unclear) hints for characterizing formal truth values for certain sentences containing bounded variables, all these "beliefs" are conventional and empty: does anybody really understand what he believes, if he believes in a "real" hierarchy of sets, a "real" domain of possible worlds, "real" space–time and so on? Just to *say* "I understand" is no criterion at all; it just begs the question. We only know (to give a simple example) what numbers are if we know how to handle number terms and arithmetical sentences. Nor do we know what a set is before we know how to name some sets and how we could talk about some classes of sets.

To exemplify this claim in more detail we briefly consider the domains of real and infinitesimal numbers. The scientific essence of the "belief" that real and infinitesimal numbers "exist" was in the beginning just the conviction that the corresponding calculi "*worked well*" – at least if applied with care.³ Logical analysis shows, then, that this belief is reasonable. After Cauchy we learned that a (name of a) real number just is a (name of a) class of convergent sequences of rational numbers. And quite recently a system of "nonstandard analysis" was developed to vindicate the way Leibniz used infinitesimals. On the basis of some set-theoretic considerations, Robinson (1966) managed to define an appropriate equivalence relation (an abstract identity) between (names of) arbitrary sequences of rational or real numbers, together with truth-conditions for elementary sentences. There results a *field* of "nonstandard real numbers" which "contains" the field of real numbers, the infinitesimals, and some "infinite" nonstandard numbers that are "bigger" than any standard real number. Even if we do not have a procedure to decide whether two (really given or merely possible) "names" name "the same" infinitesimal or not, the constitution of nonstandard numbers does not rely at all on vague "intuitions" about infinitesimals. Instead, it establishes sufficiently clear truth-conditions for certain sentences, especially equalities. The similarity to Eudoxus' definition of proportions is obvious, at least if we neglect the more basic question of how domains of sets of natural numbers are constituted such that they are "big enough" for classical, and by the same token nonstandard, analysis.

Without a constitutive definition of truth-conditions for equalities and other "possible sentences" ("propositions") it would not make sense at all to talk about abstract objects, about the truth or falsity of propositions, or about possible proofs of sentences containing names or bounded variables "in" the domain in question. The "semantical" concept of truth is the conceptual basis of deductive proofs in classical first-order axiomatic systems too; such deductions should be viewed just as technical means for deducing sentence-forms that express true statements in any semantically well-formed domain where the formal axioms express true statements.⁴ A deduction not interpreted in this way cannot show more than the following: the deductive rules of the calculus lead from formal premises (beginnings) to the end-formula. To know just this is not to know the meaning or significance of the deduction-game.

V. WHAT ARE IDEAL GEOMETRICAL FORMS?

The main logical problem of elementary synthetic geometry lies in rightly understanding the step from the "descriptive" talk about real figures or bodies to the domain of "abstract" objects like the ideal geometrical forms (the ideal points, lines, and surfaces). In geometry one seems to prove, as Plato very often stresses, true statements about ideal forms, which as such are not "observable," not "empirical things." What does this mean? How do we have to understand our talk of such forms? How do we "name" or "describe" them? What is the connection between the ideal forms we talk about in synthetic geometry and the real shapes of real physical bodies? What does it mean to say that the latter are only (incomplete) "pictures" of the former?

To obtain a satisfying answer to such questions we have to proceed as we do in defining real or infinitesimal numbers: we have to describe what counts as a (syntactically and semantically) well-formed "name" of such an ideal form, what are the identity conditions of two such names, and what the truth-conditions of the elementary sentences containing such names are. This would be quite easy if we were allowed to use the truth-conditions of analytic geometry. But then we would not see why certain arithmetical or set-theoretic statements can be viewed as *geometrical* ones or why a deduction of a formula in the chosen axiomatic system can be viewed as a proof of a *geometrically true* statement. Hence, our main problem will be to describe the *concept of truth* of sentences in (ideal) elementary synthetic geometry *independent* of *analytic* geometry, only on the basis of *geometrical demonstrations*. Only then can we see that the isomorphism between the Pythagorean (or Euclidean) *lengths* and the subspace of the Pythagorean (or Euclidean) real *numbers* really provides a translation of synthetic geometry into analytic geometry.

In the following, we focus on the concepts of *construction* and *demonstration* in elementary synthetic *plane* geometry. As we know, we can read descriptions of geometrical constructions in school geometry as *instructions*, which tell us how, step by step, to draw a picture on a (sufficiently) plane surface using the allowed means, namely arbitrary long rules with a unit length and perhaps a circle too. To make the form of such instructions a little more perspicuous we may *norm* them, for example, by the following rules:[5]

PROOF AND KNOWLEDGE IN MATHEMATICS

(0) Any construction begins with the drawing of a coordinate system, i.e. of three points P_0, P_1, P_2, where the distance between P_0 and P_1 (or P_2) is just the unit length and the clockwise directed angle $\measuredangle P_2 P_0 P_1$ in P_0 is a right one. Then there can follow a combination of the following "elementary" instructions.

(1) If the points P_i, P_j, P_k are (supposed to be) constructed (where P_k may be "equal" to P_i or to P_j), draw an orthogonal line g through P_k on the straight line $g(P_i P_j)$; call the intersection of g and $g(P_k P_l)$ "P_m" and construct on g the points P_{m+1} and P_{m+2} at a distance of the unit length from P_m. In order to make the instruction unique, here and in the following m should be the smallest number not used in a name of a point before and P_{m+1} should lie "above" P_{m+2} or, if both points lie on a line parallel to $g(P_0 P_1)$, P_{m+1} should be to the right of P_{m+2} – where the coordinate system says what "above" and "right" mean.

(2) If P_k, P_l, P_i, P_j are constructed, draw or produce the straight lines $g(P_k P_l)$ and $g(P_i P_j)$ and call their intersection point "P_m".

Formal instructions ("terms") t can be defined as sequences of elementary instructions. If we write $[P_0 P_1 P_2]$ for instruction (0), t * $[P_i P_j P_k :: P_m, P_{m+1}, P_{m+2}]$ for the continuation of an instruction t by instruction (1) and t * $[P_k P_l, P_i P_j : P_m]$ for the continuation of an instruction t by instruction (2), then an easy example of an instruction t is the following: $[P_0 P_1 P_2]$ * $[P_0 P_1 P_1 :: P_3 P_4 P_5]$ * $[P_0 P_2 P_2 :: P_6 P_7 P_8]$ * $[P_4 P_5, P_7 P_8 : P_9]$. The drawing on page 145 shows how the points must be named.

All points or lines constructed will have names of the form ${}^t P_k$ or $g({}^t P_i {}^t P_j)$, if we use the upper index t to refer to the instruction t. But lines or points in a drawing can have several names. It can happen that the points P_i and P_j or the lines $g(P_i P_j)$ and $g(P_k P_l)$ cannot be distinguished in a drawing, and it can happen that two lines do not intersect, such that in a complex formal instruction t some elementary instructions cannot be carried out. Obviously, we cannot decide just by looking at the formal instruction term t whether it is "realizable" or not. For this we have to check if the names of points P_i, P_j in the names $g(P_i P_j)$ of straight lines or circles are *different*, that is, if they *can be seen* as different in a

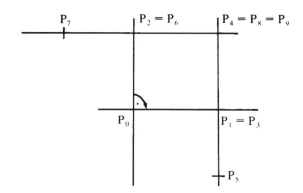

corresponding drawing, or if the production of two lines (or of one line and a circle) *actually intersect* in a drawing. We do not know this *a priori*; genuine attempts to follow an instruction t must be considered as part of the criteria for the realizability of t.

In some sense, we now see that Kant's talk about "*pure intuition*," unclear as it may be at first sight, can (or better, should) be interpreted as expressing a peculiar feature of geometrical demonstrations (of constructibility): they necessarily make use of *real intuition*, that is, of observation. With a bad drawing, however, we certainly cannot "see" that the points P_i and P_j *should be* different (or equal). Hence, we not only have to *look* at a drawing and compare it with the instruction, but we have to use *other criteria* to decide whether a picture drawn after an instruction is a "good" one or not. What criteria do we use here? Well, it would be arbitrary to let some system of axiomatic or analytic geometry decide what "should be" constructible, just as it would be arbitrary to decide it on the grounds of mere introspection or by some alleged "inner sense" – whatever Plato or Kant, Brouwer or Hilbert, or other philosophers or mathematicians might have thought about this. Using axiomatic geometry without further foundations is as conventional and dogmatic as mentalism and platonistic belief. They all resign from further attempts to understand our practice of geometrical talk and proof.

Now, what does it mean to say that certain identities or differences of angles or lengths or points, which cannot actually be noticed in a given drawing, "*should be*" distinguished in a "good" drawing? At least an important part of this question would be answered if we knew that we could or should view *any instruction* t as being realizable or not realizable independently of the unit length we start with. If we write A(t) to say that the instruction t can be realized, our question takes the following form. How are the *truth-conditions* for sentences of the form A(t) fixed, and why are they independent of an actual reference-relation between the names P_i of points occurring in t and a *particular attempt* to realize t in a special drawing?

To answer this question we make use of the theorem of Desargues. If we change the unit length on the x axis, we arrive at two pictures which are (or should be) central projections of each other. Hence, we may always choose an appropriate (big or small) size for a (part of an) instruction *to see* if it can be realized or not. The fact that we can use architectural blueprints to "predict" how a bigger or smaller realization of the same form (instruction) will look rests just on this basis.

If we assume that A(t) is "necessarily" true or false, independently of a "contingent" success or failure to realize t, all properties of ideal geometrical forms can be defined on the basis of A(t). If we say, for example, that in a form t two points P_i and P_k are the same, we do not talk about a *special* drawing: they are the same if they cannot be distinguished in *any* "good" picture, that is, if the further construction of a straight line between them *cannot* "in principle" be carried out. Thus we arrive at truth-conditions for sentences about "ideal points" and "straight lines" *named in* the instruction terms. We can even identify points named in a term t by a name ^tP_i with points named in another term t′ by the name $^{t'}P_j$, if we use the following condition: $^tP_i = {}^{t'}P_j$ is true iff in any "common extension" of the instructions – that is, in a drawing, where first the instructions of t (or t′) and then those of t′ (or t) are carried out – the points are the same. If we formulate this completely, we arrive at a logical constitution of an infinite (two-dimensional) "point-space" on a pure geometrical base. In the end this space can be shown to be isomorphic to the Euclidean vector-space of pairs of Pythagorean (or, if circles are used, Euclidean) numbers.

It is precisely the assumption that the truth-value of A(t) is independent of a particular drawing (of course, considering only "good" ones) that leads to a definition of "ideal" truth-values for geometrical sentences.[6] This assumption is the basis of any talk about "ideal" entities of elementary synthetic (plane) geometry; it is no "logical truth," however, since it must be "demonstrated" on the ground of pragmatic criteria or postulates used to judge whether an attempted realization of an instruction is a sufficiently good one, or whether it represents or pictures a corresponding abstract form or not. The proposals of so-called constructive geometry[7] to understand geometrical forms as *procedures, constructions, plans*, or *norms* of *forming* real bodies or surfaces were entirely right, if they only made sufficiently clear that and how we (have to) use *real experience* to distinguish between *realizable* instructions ("constructions," "forms") and nonrealizable instructions.

VI. PLANE SURFACES AND RECTANGULAR SOLIDS

There are many criteria or postulates we use when we judge whether an attempt to follow an instruction t is a "good" one, whether it is a "construction" or not. We have to consider the precision of the drawing and the quality of the "instruments" used, for example, if the surfaces of a rule are (sufficiently) plane, the edges (sufficiently) straight, the angles (sufficiently) orthogonal. On such grounds we judge whether a success or failure to realize t is "contingent" (due to a bad drawing) or if it is "necessary."

At first sight, the following condition for a surface to be (sufficiently) plane seems to be adequate and complete. Strictly speaking it is not; hence we use an index i (for "incomplete") and say that a surface is "plane$_i$," if this condition is fulfilled: if we mark any point on it and on any other plane$_i$ surface of another body, the two surfaces can always be brought into contact at the marked points such that there are no "empty" or "hole" spaces between the surfaces – if we focus only on places where the surfaces do not "overlap" and if we neglect possible places *outside* the plane surface which may inhibit our bringing the bodies into contact at the marked points.[8] It is rather obvious how further details might be added to make the *formulation* of our criterion even for tough-minded formalists more or less unique. We might say, perhaps, what counts as a brim or edge of a surface, or that we should be able to turn a plane$_i$ surface around, keeping the touching point with another plane$_i$ surface fixed.

These conditions for the class of plane$_i$ surfaces seem to be more or less equivalent to the "definition" of plane$_i$ surfaces as the (hoped) "outcome" of a well-known practical "construction procedure," which Hugo Dingler called "Drei-Platten-Verfahren": if we abrade and polish the surfaces of (at least three) bodies by "rubbing" and "grinding" them on each other in turn, then *in practice* we usually arrive at more or less sufficiently plane$_i$ surfaces in the above sense. But is there really a "proof" for the observed and expected fact that the Drei-Platten-Verfahren "defines" one unique class of plane$_i$ surfaces? If semantic rules are prefixed for the use of language in some context of argumentation, then at least a *logical* proof is possible. And there is no (mere) language-rule whatsoever which could tell us that two pairs of three bodies treated by the Drei-Platten-Verfahren will fit together in the way described. One might say that if this were not the case some part of the surfaces "must" be convex or concave, which would be "impossible" if we applied the grinding method thoroughly enough. But such a comment itself is based on our knowledge of what happens in the grinding procedure – and of what properties (locally) convex and concave surfaces have.

A similar problem appears if we look at the following definitions of straight$_i$ lines and orthogonality$_i$. We know what it means to lay a (transparent and thin) foil smoothly on a flat surface. The *folding* of such a foil can be viewed as an "operation" defining a "mapping" of one "half" of the flat surface into another "half." Such an operation associates possible (marked) points or lines or figures or the one half with other points or lines or figures of the other half. Using this "mapping" one could characterize a straight$_i$ line as a possible "folding line," which remains the same under the folding; a (half-)line could be said to be orthogonal$_i$ to a (straight) folding line if and only if we get a straight$_i$ line when we take it together with its "image" under the folding.[9]

Here, too, the definition must be shown to be unique, that is, the following should hold: if we fold a surface at an orthogonal$_i$ line of a folding line, one "half" of the folding line is mapped onto the other. To "prove" this, however, we rely heavily on the fact that a folding maps all the points on the left-hand side of a half-line h orthogonal$_i$ to the folding line g to the left-hand side of the image of h under the folding. Such a proof obviously uses practical knowledge, and so the consequence does not follow "analytically" from the premises by mere formal or semantic rules. I do not want to say that "proofs" like this are without any value. But they are not pure logical proofs, since they still rest on very general experience.

The full concepts of a plane surface, a straight line, and orthogonality all hang together, if we analyze the (holistic) concept of a rectangular solid, or better, the postulates we use to judge if a body has this form. It would not suffice to check if the solid has six "plane$_i$" surfaces forming quadrilaterals, each of which has three "orthogonal$_i$" angles, if "planeness$_i$" and "orthogonality$_i$" were defined independently of the postulates for rectangular solids. At least in practical circumstances it is often wise to check a further condition, namely if *all* the angles of the quadrilaterals really are *equal*, that is, if *copies* of our rectangular solid *locally* fit in all possible ways into the corresponding "*corners.*" If we were to notice differences, we would certainly suspect that either the surfaces were not plane enough, or that we did not check the orthogonality of the angles with the precision needed, or that the body as a whole was not solid enough, that it changed its form when moved. Of course, in practice we assume that if a surface is plane$_i$ and if three angles of its straight$_i$ sides are orthogonal$_i$, then the fourth angle will be orthogonal$_i$ too. This assumption, however, cannot be "proved," though there have been many attempts at such a proof in the history of geometry.

We can use nonstandard analysis to show why or in what sense the concepts of a plane$_i$ surface, a straight$_i$ line, and orthogonality$_i$ really are incomplete.

In a nonstandard extension $(R^3)^*$ of the three-dimensional number space R^3 there are analytical spheres having infinite radii. Let M be the center of such a sphere S with radius r, where r is *finite* if we look at it in nonstandard analysis, *infinite* if we look at it in standard arithmetic. The set E of points on that sphere at only standard-finite distance from a fixed point P on S form a "surface" (in an analytical sense). This surface cannot be distinguished by standard distances from any tangential plane touching E at any point – if we neglect the "overlappings." Each circle on S with center M and radius r may be viewed as a "folding line," if we map one-half of the sphere in an appropriate way onto the other. If we interpret the formulations correspondingly, we arrive at just the standard concept of the orthogonality of such "folding lines" on spheres which is defined by the orthogonality of the corresponding tangent lines. If we *do not neglect* infinitesimal differences, then, certainly, *there are no rec-*

tangulars in E, since there are no rectangulars in our sense on a sphere. On the other hand, E *is* plane$_i$ if in our "fitting" condition for planes$_i$ we neglect infinitesimal distances. By this, we get a semantic interpretation of the relation of "fitting" of surfaces, the function of "folding" of surfaces, and the concepts of "plane$_i$ surface" and "orthogonality" in an abstract analytic model, such that the defining sentences (of Lorenzen's) are satisfied, if we take them literally – as we must do if we want to apply formal logic. Hence, the assumption that on every plane$_i$ we can "construct" rectangulars by constructing four lines a, b, c, d such that a and b, b and c, and c and d are orthogonal$_i$ does not follow logically from the defining properties formulated in constructivist geometry. Instead, it must be viewed as part of the defining properties of the full concept of a plane we work with in elementary synthetic geometry.

We use still other criteria to check the quality of a rectangular solid, for example the following. Given any pair of rectangular solids S_1 and S_2, we should be able to build with the help of copies of S_1 a rectangular solid bigger than S_2. This is a special form of the more general postulate that geometrical sizes (volumes, surfaces, straight lines, angles) form Arcihmedean linear orderings. Furthermore, we assume that rectangular solids can be put together from two equal (congruent) "rectangular wedges," the diagonal halves of rectangular solids. This postulate is a pragmatic substitute for the famous parallel postulate. With its help we can immediately demonstrate that plane surfaces or straight lines do not intersect if and only if both stand orthogonally on a plane surface or a straight line. (Note that we do not intend to give a complete list of criteria here, nor do we present any geometrical demonstration in all details.)

The existence of analytic models of rectangular solids together with analytic representations of movements and fitting conditions in the number space R^3 proves at most the *formal consistency* of our fuller list of criteria for rectangular solids. Pragmatically it is just a fact that there are "good" rectangular solids coming in a wide range of size and precision. And their realizability in a large range of size and precision is part of the *complete* concept of planeness and orthogonality, and of the very concept of truth and proof in elementary synthetic geometry. The special ways we arrive at good rectangular solids and wedges does not matter at all.

In mathematics we do not assume a finite end to the series of (names of) natural numbers; and we do not assume a maximum or minimum size of "possible rectangular solids." Idealization in mathematics is just this "abstraction" from concrete empirical boundaries. Such idealizations are necessary for the easy and invariant way of talking in mathematics about the set of numbers and their properties or about the abstract domain of geometrical forms. On the other hand, idealizations of this sort "create" the only concepts of infinity we really understand: infinities exist just as mathematical ways of speaking about procedures without prefixed limits, as it was claimed by Kant.

On the basis of the idealizing assumption that there is an unbounded number of ways of constructing rectangular wedges fulfilling the pragmatic criteria, that is coming in "any" size and precision we wish or need, we can demonstrate the central theorems of elementary synthetic plane geometry: the theorems of Thales and Pythagoras, the theorem of Desargues and, hence, the invariance of the truth-conditions of A(t). This is done by showing why certain "movements" of bodies (wedges) or surfaces (triangles) are (and should be) possible, or why it is (or should be) possible to build up a certain figure out of certain parts. These possibilities are "necessary" given the assumption that the "material" is solid and well-formed enough. *This*, not some mystic "pure intuition," is the concept of proof in synthetic geometry leading to the concept of truth in Euclidean geometry and to the isomorphisms between the number spaces P^2 (or E^2) of pairs of Pythagorean (or Euclidean) numbers and the points on a plane constructible by Pythagorean (or Euclidean) means.

VII. WHY ARE GEOMETRICAL TRUTHS SYNTHETIC *A PRIORI*?

In P^2 or E^2 we can certainly prove truths in a purely "semantic" way: the truth-values of the sentences "about" these number spaces are defined in the framework of an algebraic extension of the rationals Q or, if one prefers to be more complicated than necessary, in the real numbers R. Our "translation" between synthetic geometry and analytic geometry shows, then, why we can use the true sentences in these spaces to check *a priori* if an instruction t can be realized or if an attempt to do so is futile or if it is a good or a bad one. Without such a basis it would be rather arbitrary and not too wise to use the truths of formal analytic systems as *a priori* criteria for judging whether it is worthwhile to try to fulfill or "exhaust" some plan or instruction. If we did not justify the invariance of A(t) on geometrical grounds, using the theory of similarity (e.g. the theorem of Desargues), or if we just said that this invariance is the basis of our talk about forms and that it is a powerful tool for geometrical proofs, our reconstruction of synthetic geometry would not be satisfactory; in fact it would not be much better than an axiomatic procedure.[10] Instead, our reconstruction shows the following. If we use certain true sentences of Euclidean geometry (be their proofs analytic or not) to control the stability of the geometrical form of a physical body or a space inside certain limiting surfaces we rely heavily on the very basic *fact* that in "middle" sizes and "middle" margins of precision sufficiently good rectangular solids and wedges or their hollow forms as standards of comparison *are actually available.*[11] If we use Euclidean geometry for describing spatial relations of bodies in microcosmic or macrocosmic dimensions, however, we (must) use a *fictional* extrapolation of our real possibilities, namely that the form of rectangular solids (or of planes and straight lines) of the appropriate size and precision *could* be realizable by some other means than by solid bodies. Since in reality such a hope could prove to be futile, it would be unwise to decide entirely *a priori* that Euclidean geometry *must* be used as a framework for *all* spatial orderings of physical phenomena.

As Kant pointed out, geometry is an example *par excellence* of what we call a system of *synthetic* (not analytic) truths, which we nevertheless can *prove a priori*. We have shown here, however, that we have to understand that "*a priori*" in a *relative*, not an *absolute*, sense: geometrical demonstrations rely on observable facts. Indeed, other proofs do also. If we were not able, for example, to distinguish figures like "a" and "b" by the help of our sense organs in an intersubjective and stable way, we certainly could neither hear nor speak nor read nor follow figurative deduction rules. Every analytic truth relies on some such facts, especially on the fact that we can use syntactic and semantic rules in a uniform way. In normal discourse, however, we do not consider facts of this sort as "empirical facts." Rather, we presuppose them. Since the presuppositions used in elementary geometry are not as simple as those just mentioned, their logical analysis is at the same time more difficult, more important, and more interesting.

VIII. MATHEMATICAL MODELS OF A BOUNDED ("FINITE") SPACE-TIME

We know nothing about space and geometrical form apart from our knowledge of the possibilities of moving bodies from one place to another, how they "fill" a space, and how they "fit" together with each other. Haptic sensation provides us with information about the possibilities of moving ourselves and about other fitting conditions; optic sensations give us geometrical information only indirectly, by producing pragmatic expectations about the "haptic" geometry of the object or space observed. Leibniz was aware of this "relativity principle" of the spatial ordering of bodies, and Kant rightly discovered that the "orientation" in space rests on direct intuition, that is, on the standpoint of the observer.

Now, there is an important consequence of our analysis of the concept of geometrical proof or demonstration. It is not too surprising that there are non-Euclidean axiomatic systems in which many theorems of Euclidean geometry still hold but the theorem about the uniqueness of parallels does not; nor should we be astonished that there could be a need to develop some special alternative mathematical framework for representing the result of *macrocosmic* (or microcosmic) spatial measurements, if we use (light-) rays and clocks to measure cosmic distances. This last possibility depends on some knowledge of the relative behavior of the rays and clocks. The *local velocity* and propagation of light, for example, is measured by solid rods (rectangular solids) and clocks. Since these measurements have stable results, we have good grounds for working with a *Euclidean* space-time "geometry" in all "middle" and small sizes. For the same reason we can use time measurements of the propagation of rays to measure lengths of middle size *indirectly*. But what could be an empirical reason for assuming that the velocity of light is *constant* in cosmic dimensions and that the rays are "*straight*" (or better, that the propagation of light around an emitting body is a real "ball")? An assumption like this seems to presuppose the (fictitious) possibility of checking the geometric form of the propagation of light by macroscopic rectangular solids. Since this is impossible, we might prefer to replace any such assumption by the *conventional stipulation* just to *call* "light rays" and/or the "paths" of movement of inertial systems "straight," and *define* cosmic distances on the basis of some time measurement of the propagation of rays. Then, however, we will have to check if the properties of *these new* concepts of "straightness" and "distance" together with the local concept of an "angle" between such "lines" have the same formal properties as the Euclidean ones we have dealt with up to this point or different ones. And if we represent light rays in geometrical drawings by straight lines – since we might have good grounds to view them as the "straightest" lines which can (or could) be realized in cosmic dimensions – we still have to be careful when we interpret such pictures. Riemannian geometry provides us with a general analytic framework in which "locally" Euclidean geometry holds and some "global" properties of "space–time" can be determined by local properties, namely the local curvature of the "straightest" lines and "flattest surfaces" in (four-dimensional) space–time. This idea goes back to Gauss, who managed to map a "global" landscape by local measurements. Doing this, we proceed in a similar way as in Euclidean geometry: empirical knowledge and some (conventional) idealizations are the basis of the mathematical space–time geometry in general relativity theory.

If the number of a length is calculated from a measured number of some time interval, our interpretation of such numbers certainly must take the properties of the time measurement into account. There, of course, we *wish* that good synchronized clocks show the same time if we move them around. We are quite good at producing such movable clocks for our *local* time measurements. The representation of time as a line just pictures this, since we are used to ordering "all normal" phenomena in a linear way. But certainly the assumption that there "might" be clocks showing the same dates (time-numbers) even if they are moved "cosmic" distances in "cosmic" velocities could be wrong – at least if we take *actual possibilities* into account and do not use the word "possible" only in the sense of "thinkable," which only means that something cannot be excluded on the ground of mere *formal* considerations. Even if we can "think" of a "possible" instantaneous comparison of local events which take place far apart from each other, there is no empirical realization of this formal (and hence mere rhetorical) talk. Indeed, it is well known that two locally synchronized clocks normally will *not* show the same date (represented by some time-number) if they are moving with great velocity away from each other and later meet again. This is a consequence of the empirical content of the famous Lorentz transformation: it correlates the time-number of inertially moved clocks with the time-numbers defined by Einstein synchronization, which "dates" events on a distant inertial system on the basis of a local measurement of the time a ray needs to go there and come back.[12] Given this background, a uniqueness proof for "absolutely" good clocks – as Janich (1985) sought to provide – can be significant only if it can show that properties of dynamic systems which can be controlled *locally* have the *global* consequences we expected from "good" clocks. But, alas, there *cannot* be a proof like this: the only thing we know is that in "normal" (local, not cosmic) situations we can fulfill our wishes and produce "good" movable clocks (special dynamic systems) allowing quite precise intersubjective datings and stable measurements of time intervals of certain typical events.

In the end it follows that constructivist protophysics fails to arrive at a pragmatic foundation of a pure and *a priori* "Euclidean kinetics" independently of physical dynamics. If we use light or other rays for transmissions of information (e.g. about the number on a clock) into far distances, we rely on some knowledge and some assumptions about the relative behavior of the propagation of light and the dynamic systems used for the local time measurement. There is no ground to claim that we *should* use Euclidean geometry as a *general* framework for "exhausting" all "good" methods of measuring "lengths" – as Dingler and Janich proposed – and that "good" clocks *must* produce a linear chronological ordering of all events, even those in cosmic distances. Of course we take an interest in *simple* ways of mathematical representations – and hence we would prefer ideal clocks to the real ones we have, and we could wish to have a realization of large "straight lines" in a Euclidean sense. But if we have a problem with fulfilling such wishes or with picturing our actual experience by a formal (mathematical) model, then it is not too wise to stick to the wish or to a traditional framework we are used to working with. Rather, it could be a challenge for mathematicians to adjust the system of representation to some new knowledge or, even more importantly, to a new method of measurement and the corresponding presuppositions. Exactly this was done in relativity theory. And, indeed, if we want to understand the conceptual achievement of Einstein's work, neither a realistic (Pythagorean) nor a mere conventionalistic interpretation will be appropriate.

NOTES

1 Cf. K. v. Fritz (1945: 242–62), Becker (1966: 71ff.) and Fowler (1987: 33ff.).
2 Cf. the modern interpretations of Aristotle: Becker (1933: Topica 158b29ff.; 1966: 102ff.) and Fowler (1987: 17f., 32f.).
3 Leibniz seemed to be quite aware of the fact that the construction of a suitable *symbolic language* or appropriate *formalism* is often crucial for the development of mathematics.
4 Not every truth (true sentence) in a semantically well-formed domain can be proved by deducing it in an axiomatic system. If we show the truth of the axioms in the domain, that is, that the domain is a *model* of the axiomatic system, we have to show that the definitional truth-conditions of the domain are fulfilled if we "interpret" the formal axioms in an appropriate way. Such an interpretation is nothing other than an attachment of truth values to the formulas following the definitional rules of the domain: it transforms the formulas into (true or false) sentences. The correctness theorem of predicate calculus shows, then, why we can call a deduction of a sentence-form a "proof" of a corresponding sentence (proposition) in any domain where the sentence-forms represented by the (formal) axioms hold get the value "true."
5 In the following, we consider only Pythagorean constructions, that is, we neglect the use of a circle. Instruction (1) is formulated in a rather complicated way in order to avoid lengthy case distinctions.
6 Lorenzen and Inhetveen call this assumption "Formprinzip," even if they formulate it differently. Cf. Lorenzen (1984a: II, §3), especially pp. 101f., and Inhetveen (1983: §II.2), especially p. 68.
7 Cf. Lorenzen (1977, 1984b), Inhetveen (1983) and Janich (1976).
8 The formulation is a short variant of the one given in the works of Inhetveen (1983: 29ff.) and Lorenzen (1984a: I, §3).
9 Once again we give a short variant of a definition proposed by Lorenzen (1984a: I, §3, 4).
10 This is the reason why I judge the reconstructions of Inhetveen and Lorenzen, subtle as they are, as still unsatisfactory.
11 Things of middle size are those we deal with in everyday life. What are "microscopic" and "macroscopic" sizes must be explained in contrast with them: we must *show*, not *define*, what belongs to the (admittedly vague) range of middle sizes.
12 Whereas Lorentz still seemed to think that the transformations named after him are based on a mere definitional convention, Einstein saw that they express an empirical truth, since they correlate the behavior of the propagation of light with that of local dynamic systems (watches) in an inertial system.

REFERENCES

Becker, O. (1933) "Eine voreudoxische Proportionionenlehre und ihre Spuren bei Aristoteles und Euklid," *Quellen und Studien zur*

Geschichte der Mathematik, Astronomie und Physik, Abteilung B: Studien 2, 311-33.

────── (1966) *Das mathematische Denken der Antike*, 2nd edn with an appendix by G. Patzig, Göttingen: Vandenhoeck & Rupprecht.

Euclid (1883-1916) *Opera omnia*, ed. J.L. Heiberg, H. Menge and M. Curtze, 9 vols, Leibzig: Teubner.

────── (1956) *The Thirteen Books of Euclid's Elements*, transl. from the text of Heiberg, with introduction and commentary by Sir Thomas L. Heath, 3 vols, 2nd revised edn, New York: Dover Publications.

Fowler, D.H. (1987) *The Mathematics of Plato's Academy. A New Reconstruction*, Oxford: Clarendon.

v. Fritz, K. (1945) "The discovery of incommensurability by Hippasus of Metapontum," *Annals of Mathematics* 48: 242-64.

Hilbert, D. (1900) *Grundlagen der Geometrie*, Stuttgart (9th edn, 1962).

Inhetveen, R. (1983) *Konstruktive Geometrie. Eine formentheoretische Begründung der euklidischen Geometrie*, Zürich (BI).

Janich, P. (1976) "Zur Protophysik des Raumes," in G. Böhme (ed.) *Protophysik, Für und wider eine konstruktive Wissenschaftstheorie der Physik*, Frankfurt/M., 83-130.

────── (1985) *Protophysics of Time*, Boston Studies in the Philosophy of Science, vol. 89, Dordrecht, Boston, MA, and Lancaster: D. Reidel.

Lorenzen, P. (1977) "Relativistische Mechanik mit klassischer Geometrie und Kinematik,' *Mathematische Zeitschrift* 155: 1-9.

────── (1984a) *Elementargeometrie. Das Fundament der Analytischen Geometrie*, Mannheim (BI).

────── (1984b) "Neue Grundlagen der Geometrie," in P. Janich (ed.) *Methodische Philosophie. Beiträge zum Begründungsproblem der exakten Wissenschaften in Auseinandersetzung mit Hugo Dingler*, Zürich (BI), 101-12.

Robinson, A. (1966) *Nonstandard Analysis*, Amsterdam: North-Holland (2nd revised edn, 1974).

6

MATHEMATICAL RIGOR IN PHYSICS

Mark Steiner

SUMMARY

Physicists for centuries have been making correct numerical predictions on the basis of nonrigorous mathematics. Their arguments, though deductively invalid, were bolstered by extra-mathematical considerations and approximation techniques. Recently, however, physicists have been flouting mathematical rigor with no pretense of before-the-fact justification. Mathematical rules for symbol manipulation are employed beyond the range of their validity. Mathematically inconsistent theories are employed. The result? The greatest accuracy in the history of physics (in the case of quantum electrodynamics, 1 part in 10 billion). How do physicists do it?

When I studied modern algebra as an undergraduate, the instructor, Professor Ellis Kolchin, contemptuously defined a physicist as "someone who thinks that a vector is an ordered triple." Roughly at the same time, Murray Gell-Mann was arguing that, really, physicists do not need to know mathematics, since the drive for abstraction and rigor characteristic of modern mathematics is only to avoid recondite counterexamples that never appear in nature.[1] While these two statements probably could not be made today – we live in the era of string theory, whose practitioners no longer know whether they are doing physics or mathematics – they do reflect an attitude by physicists to both mathematical abstraction and mathematical rigor.

In *mathematics*, mathematical rigor[2] ends disputes about whether a proof has, or has not, been presented – or whether a certain concept or theory is consistent. More subtly, mathematical rigor forces mathematicians to make distinctions otherwise left vague (such as the distinction between continuity and uniform continuity), thus enriching mathematics. Nevertheless, mathematicians have not always presented rigorous proofs – even when they were capable of doing so. As Imre Lakatos shows,[3] premature rigor can stultify the growth of mathematics. There are other epistemic values in mathematics than truth (whatever mathematical truth is, exactly) – such as, for example, explanatory adequacy – and, at concrete stages in the development of a theory, these values may conflict.

For example,[4] Euler gave a nonrigorous proof that, for "polyhedra," $V - E + F = 2$, the variables being, respectively, the number of vertices, edges, and faces of the "solid." But the modern point of view is that Euler's theorem is "really" about the topology of surfaces, not the geometry of solids; a theorem in topology, not geometry.[5] It would, indeed, have been possible to concoct a definition of "polyhedron" for which the proof became rigorous and exceptionless; but if that point of view had been adopted, the modern, deeper, point of view might not have been discovered.[6]

The utility of mathematical rigor in *physics* follows from the primary physical tasks of mathematical reasoning: deriving new laws from old, and predicting experimental consequences from theory. Thus, Newton derived[7] the inverse square law of gravitation from Kepler's laws; and Einstein calculated the precession of the orbit of Mercury from general relativity. Now, mathematical rigor ensures deductive validity – if the premises are true, the conclusion will be true.[8] However, in physics there is an independent check on our conclusions – experiment and observation – and the need for rigor is, correspondingly, less critical than it is in mathematics.

Furthermore, even nonrigorous reasoning may be reliable when it is supplemented by intuitive, or empirical, considerations. Euclidean geometry is not completely rigorous – for example, it contains no discussion of the crucial concepts of "inside" and "outside." In high school geometry and in physics, therefore, we rely upon our visual intuition, or perhaps careful drawings, to decide whether various lines meet inside or outside a triangle. Naturally, we can be deceived – as in the well-known "proof" that every triangle is equilateral – but we can be deceived also in applying deductive logic. Another example: physicists blithely assume that a mathematical function arising in physics is continuous and differentiable, until proved otherwise. There is a basis in experience (if not reason) for this working hypothesis. Even in quantum mechanics, the discontinuities occur, not so much in the mathematical formalism of the theory (classical analysis), but in its interpretation.

Even when intuition fails us, as in the case of the infinitely large or the infinitely small, so that lacking rigor we cannot even formulate our conclusions consistently, much less prove them, the physicist can sometimes rely on the intuition that these conclusions express at least approximate truths. An example of this is the calculus, not put on a consistent basis until the nineteenth century. Bishop Berkeley, we recall, attacked the Newtonian "fluxions" (time derivatives) as "ghosts of departed quantities." Newton was accused of regarding, inconsistently, an increment in a variable simultaneously as something and nothing. Berkeley aimed to undermine Newton's intellectual authority by showing that, compared with calculus, theology looks positively coherent. He did not, however, argue that the calculus does not work in physics. Take, for example, the function $y = x^2$. If we add a small increment h, the difference quotient $(f(x + h) - f(x))/h$ is $2x + h$, which we can make as close to $2x$ as we like. Thus the value $2x$ can be used, instead of $2x + h$, in calculations.

Even in the twentieth century, physicists persist in thinking of the derivative intuitively, as a quotient of "infinitesimals." They think of integration as a summation of infinitesimal area (or volume) elements. In line integration, they think of a curve as the union of infinitesimal straight line segments, etc. Similarly, they talk of infinitesimal rotations, rather than tangent spaces of Lie groups. It was only recently that even this talk of infinitesimals was legitimized rigorously by Robinson's work, and even now, so far as I know, nonstandard analysis has not yet been brought to bear on the infinitesimal rotations.

Following in Newton's footsteps, twentieth-century physicists have invented their own mathematical "limit" concepts, whose consistent formulation came later, if at all. Dirac introduced the Dirac δ-function, in analogy with the Kronecker δ-function. The latter is a function usually defined on the integers: $\delta(i, j)$ is 1 or 0 depending upon whether $i = j$ or not. The Dirac function, intended to be used under the integral sign (not by itself) in describing systems with continuous energy spectra, is defined on the reals, and is infinite when $i = j$ – yet its integral is 1. This idea was formalized subsequently.

A concept still looking for its Weierstrass, to say nothing of its Robinson, is the Feynman path integral. An ordinary integral, as used in physics, is a sort of weighted sum of the values of a function over a (closed subset of) a line, a plane, or a higher-dimensional space. The Feynman integral is the weighted sum of a function over an infinite dimensional space![9] No one knows yet how to define in general an integral over such a space.

Nevertheless, the idea has such intuitive validity – to the physicist[10] – that the Feynman integral is a basic tool in formulating quantum field theory, and in predicting the results of experiments to great accuracy. What is more, in a small number of specific cases, the Feynman integral can be rewritten as an ordinary integral.[11] As Manin puts it, "Most problems of quantum field theory can be thought of as problems of finding a correct definition and a computation method for some Feynman [sic] path integral. From a mathematician's viewpoint, almost every such computation is in fact a half-baked and *ad hoc* definition, but a readiness to work heuristically with such *a priori* undefined expressions ... is necessary in this domain."[12]

Nevertheless, as the Talmud observes, "one sin leads to another": as we shall see, the Feynman integral, mathematically questionable in itself, leads to further violations of mathematical rigor when it is used in calculations.

Let us sum up the discussion till now:

(a) Physicists have been flouting mathematical rigor for centuries.
(b) Physicists have used, even invented, quasi-mathematical concepts, whose consistent formulation was only subsequently, if at all, presented by mathematicians.
(c) These violations of rigor have not prevented accurate calculation of observable facts.
(d) There is nothing mysterious in these successes, since the reliability of the reasoning was buttressed by intuition, or by induction. Furthermore, it was often clear, even deductively clear, that unformalized mathematics could be used at least to approximate the facts.

I now claim that twentieth-century physics, aside from flouting rigor in the traditional ways, has introduced a disturbing new type of rigor violation. Namely, contemporary physicists employ pseudo-mathematical reasoning unbuttressed by plausible arguments – and still get good results! To put it another way, the symbolic manipulations performed by present-day physicists bear only formal analogy to mathematics.

By "formal analogy," here, I mean that the analogy is to the syntax of the mathematical expressions, not to their content.[13] To give a classical example of formal analogy, consider Euler's manipulation of power series in formal analogy to polynomials, as though a power series could be considered a polynomial of "infinite degree," factored as an infinite product (as a polynomial can be factored as a finite product of linear factors), etc. Euler had no mathematical justification for this. In some cases, the power series themselves diverged, so that the analogy to polynomials was even more tenuous.[14]

Modern physicists resemble Euler not only in that they violate rigor, but in the details of such violation. They, like Euler, manipulative divergent power series, without standard mathematical justification. Thus, I conclude, their after-the-fact success is quite remarkable, and presents a philosophical problem. In what follows I shall present two examples of this type of violation.

The first example relies on a generalization of the notion of a power series, namely, that of an analytic function. An analytic function (of a complex variable) is a function that looks, locally, like a power series.[15]

Suppose we have a function $f(z)$, analytic in a region R. Then there is at most one way to continue (extend) the function $f(z)$ so that it is analytic in a larger, containing region S. For a spectacular example, consider the function (the Riemann ζ function)

$$\zeta(s) = 1/1^s + 1/2^s + 1/3^s + \ldots$$

where s is a positive integer; $s = 1$ gives the (divergent) harmonic series; $s = 2$ gives the sum of the reciprocals of the squares; summed by Euler as $\pi^2/6$, etc. In fact this series converges for every real value of $s > 1$ – indeed, for every complex number $s = x + iy$ in the half plane $x > 1$. It is not hard to show that this function is analytic in that half plane. Elsewhere in the plane, the infinite series does not even converge. Yet it is possible to find a formula (involving complex integrals, that is, not looking at all like the series $\zeta(s)$) that agrees with $\zeta(s)$ in the half plane $x > 1$, yet defines an analytic function everywhere in the complex plane except for the one point on the real axis, $s = 1$. This continuation must therefore be unique, so that the extended function can be called the Riemann ζ function.

Suppose, now, that we have a function $f(n)$ defined on the natural numbers – for example $n!$, defined as $1 \times 2 \times 3 \times \ldots \times$

$(n - 1) \times n$. Of course f, so defined, is not an analytic function, but there may be infinitely many functions g that extend f and are analytic throughout a region containing the natural numbers. In fact, the "Γ function" is a mathematically interesting function that almost[16] does just this for the function $n!$ Suppose we have, say, a power series, defining, at and near the number m, any one such function g (i.e. one that is analytic in a region containing the natural numbers and which agrees with f on the natural numbers). Then, by continuing the function g to a region containing n, we could actually calculate the value $f(n)$. It would be ludicrous to calculate 11! by calculating the Γ function as a power series near, say, 13, and then continuing down to 11; we can calculate 11! by direct multiplication. Yet, in physics, direct calculation of a function may not be possible, as in the following.

Problem: to calculate the critical exponent of a ferromagnet (i.e. how fast it loses its magnetism above its "critical temperature"). This problem in thermodynamics proves impossible to solve, even by approximation methods, from first principles. Instead, one calculates the critical exponent of a *four-dimensional magnet*, and *analytically continues* the solution from dimension four to dimension three!

This procedure, however bizarre intuitively, would be mathematically rigorous if we could make the following (drastically oversimplified) assumptions.

(a) The function $f(n)$, which gives the critical exponent of a magnet of dimension n (and which is well defined), can be generalized to an analytic function $f(d)$, at least in a connected region containing both 3 and 4, by (among other things) replacing the factorial function by the Γ function wherever the former occurs in the mathematical expression for $f(n)$.[17]
(b) Around $d = 4$, we can express the function $f(d)$ by a power series $s(d)$, whose radius of convergence is at least 1.

Unfortunately, assumption (a) is not known to be true; and, if s is the actual power series used in calculation, (b) is definitely false. For the series used to approximate the hypothetical function f, around $d = 4$, is known not to converge. We have, then, a calculation that violates all canons of rigor.

Intuition is of no help here; what *intuitive* reason is there that the critical exponent of a magnet should be an analytic function of dimension? Nor is there any intuitive reason why a conconvergent series should be "continuable"[18] from 4 to 3. Finally, since the mathematical concepts in this example do not correspond to empirical concepts – they cannot be "reflected" downward – there is no good reason why the experience of the past should be relevant here.[19] There are, indeed, cases in which dimension was successfully regarded as a continuous parameter. In quantum electrodynamics, calculations were made in dimension "four minus epsilon" and then continued to four.[20] But there is no – known – *physical* analogy between these cases. Thus the analogy is only formal, that is, physically irrelevant.

Of course, it could still turn out that dimension really *is* a continuous parameter, that is, that the analogy is physical, and that there is a unified explanation for the variegated success of dimensional regularization. This would not solve the *epistemic* puzzle, however, since this explanation was not known at the time the analogy was made.

Here is another example of crass violation of mathematical rigor, this time from quantum electrodynamics. According to Dirac's theory, the magnetic moment of the electron is 2. Field theory introduces a correction to this, caused by "vacuum polarization," that is, the spontaneous creation of charged particles and anti-particles in the vicinity of the electron. The calculation of this perturbation generates an infinite power series (call it S), with infinitely many divergent integrals as coefficients![21]

Surprisingly, however, there is a way to get numbers out of this seemingly inconsistent formalism. Given any natural number n, there is a method for replacing the first n coefficients by finite, that is, convergent, integrals. The method is called "renormalization," and, for the benefit of the lay public (who are unaware of the scandalous doings of modern physicists), I will expose it here.

Express such magnitudes as electric charge by formal power series of the same type: that is, by power series whose coefficients are themselves divergent integrals taken over all wavelengths. Thus, in our original formal power series, we substitute others, ignoring, of course, terms beyond the nth. Rearranging terms (manipulations which, by the way, require mathematical justification), we can find that the divergences cancel each other out; all integrals are finite. Now, in theory, we can do this for every n, producing a new infinite series S^* whose coefficients are defined. In practice, calculating even the first four coefficients is a very arduous task, so that almost all of the series must be thrown away.

Let us look at what is meant by divergences "cancelling" one another. We cannot literally subtract infinity from infinity. How, then, can divergences "cancel"? Instead, all integrals are made functions of a "cutoff": we integrate over all wavelengths up to an unspecified finite limit L, so that they are all meaningful functions of L. In the case of meaningful integrals, of course, the difference between integrals is equal to the integral of the difference. It is these difference integrals that converge when we finally allow L to go to infinity.

Now, this procedure would be perfectly rigorous on the following two assumptions:[22]

(a) There is a "natural" cutoff L^*. That is, in reality, in making calculations in quantum electrodynamics we need not take into account frequencies higher than L^*.
(b) The infinite series S^*, generated by renormalizing infinitely many times (as n goes to infinity), converges. (So far we have established only that the *coefficients* of S^* converge – this says nothing about the series.)

What is the status of these assumptions? We can dispose of (b) rapidly: there is no evidence that S^* converges.

As for (a), the situation is more complex. A natural cutoff for frequencies would follow from an assumption found already in David Hume – that, despite what mathematicians from Thales on say, there is a minimum "atomic" distance.[23] However, though this proposal rears its head every now and then in the world of science, so far no one has successfully propounded a physics with a minimum distance. One of the problems facing any such theory is that it apparently contradicts the theory of relativity, since any cutoff would have to be relative to the observer.[24]

Granting the evidence available today, and certainly that available to propounders of quantum electrodynamics, the renormalization procedure alluded to here is not rigorous. In fact, though the substitution of series for magnitudes like charge or mass can be justified,[25] the rest of the procedure, with its unsupported assumptions of cutoff and convergence, cannot. Thus, although renormalization provides us with defined quantities the argument that these quantities should be the ones we were trying to predict is certainly lacking rigor. It is perhaps ironic, then, that the most astonishing accuracy in the history of physics was achieved by renormalizing quantum electrodynamics – in predicting, as above, the magnetic moment of the electron. The accuracy of the prediction is better than 1 part in 10 BILLION.[26] It is fitting the Feynman, who taught physicists how to do such calculations, characterized the role of mathematics (in general!) in science as "following rules that have nothing to do with the original thing" (Feynman 1967: 171).

What we have here, really, is pseudo-mathematics – which, indeed, may be replaced some day by rigorous reasoning. However, the standard apologia – that when physicists use "bad" mathematics they actually intend these rigorous replacements – is credible only for reasoning that is backed, unlike renormalization, by intuition or experience.

There are philosophical morals here.

Philosophers have held up an idealized image of mathematics and mathematical proof. Reasoning, at least mathematical reasoning, in science should be formalizable in a consistent, perhaps first-order, system. Thus, the various models of scientific knowledge put forward in the writings of such philosophers as W.V. Quine, Carl Hempel, and Isaac Levy, however much they may differ among themselves, assume that a consistent formal system of mathematics and logic underlies science. In fact, these writers go further, and assume (though for different reasons) that the standard mathematical theories used in science are not consistent, but true. Yet it is doubtful whether we can attribute to *today's* physicists even a consistent mathematics.

Only Wittgenstein treats both logic and mathematics as "forms of life" among others, and looks with suspicion upon the "superstitious dread and veneration by mathematicians in face of contradiction."[27] In a debate with A.M. Turing,[28] Wittgenstein maintained that "hidden" contradictions might not interfere with the applicability of a mathematical system in engineering – against Turing, who argued that an unnoticed contradiction might result in a bridge falling down. The examples discussed here show that Wittgenstein's point of view has much to recommend it.[29]

These reflections lead to some speculations. That the rules of mathematics lead to the right results might seem to some a trivial matter. After all, these rules are either truths or derived from truths – why should not truths lead to truths? We have seen, however, that mathematical rules (for the manipulation of symbols) can lead to correct results outside the domain of their validity.[30] Surely there is something to explain here. Physicists manipulate symbols in ways that externally resemble rigorous mathematics. The success of such behavior, even if only partial, should be cause for surprise if not wonder – since it has led to such accurate predictions (and thus differs from astrology, for example, where partial "success," I take it, need not be explained at all). For what the success of this behavior suggests is some kind of correspondence between the laws of thought[31] and the Universe.[32] Modern naturalism, it is true, allows only those connections between the brain and the world that can be explained by natural selection. But while the analogical behavior of physicists, like analogical behavior in general, is understandable, and certainly has had survival value, the predilection for analogy, in some contexts, can lead to superstition and magic. Magic often involves transferring properties of the world to words (e.g. attempting to do away with one's enemies by erasing their names), that is, it involves inappropriate analogies. An evolutionary analysis might well lead one to classify contemporary physics, with its reliance on symbolic analogies, with magical behavior!

Magic or not, science works. One cannot blame physicists for using procedures that effect such accurate predictions. They can always hope that their calculations will some day be embedded in rigorous mathematics. Philosophers, however, are left with the epistemic problem: how is it that procedures suggested by irrelevant, formal analogies so often turn out later to be justified? I commend the problem to the reader.[33]

NOTES

1 See Gell-Mann (1987: 511).
2 As a first approximation, a mathematical proof will be said to be rigorous if it is presented in such a way that an imaginary logician, otherwise unskilled in mathematics, could nevertheless transcribe the proof in a first-order predicate calculus. See Steiner (1975).
3 What I regard as most important in Lakatos – the reader should be warned – is not necessarily what Lakatos emphasizes in *Proofs and Refutations* (1976). See my "The philosophy of mathematics of Imre Lakatos" (1983).
4 This example is dealt with at great length in Lakatos (1976).
5 For example, no matter how we divide a two-dimensional sphere – or any surface topologically equivalent to a sphere – into "triangles," $V - E + F = 2$. If we do the same for a torus (inner tube of a tire), the Euler sum is zero, showing the topological inequivalence of the

torus with the sphere. As Lakatos argues, the "counterexamples" to the theorem are "accepted" as on a par with the theorem.

6 See also Philip Kitcher (1981) and the concluding chapter of *The Nature of Mathematical Knowledge* (1983).
7 The word derivation here means a plausible argument which allows a physicist to "write down" an equation.
8 This sentence, and what follows, remain correct when "empirical adequacy," in the sense of Bas C. Van Fraassen (1980: ch. 3), is substituted for the word "truth." This essay is intended to be neutral on the disputed question of scientific realism.
9 Given some physical system, we think of all the possible paths through space – time leading from state 1 to state 2, and give each one of these paths a "weight." Then we take the weighted sum over an entire space of possible paths to get the total "weight" of the transition from state 1 to state 2. This space of possible paths is of infinite dimension as each path is itself a continuous function of a real variable.
10 For example, consider the Fermat principle that light chooses the swiftest path, consistent with the constraints of its environment, to get from one place to another. This idea is teleological: how does the photon know in advance which is the optimum path until it gets there? In the Feynman integral formulation, in fact, light uses every possible path, but the paths that deviate from the Fermat path cancel one another, leaving the Fermat path as most probable.
11 What is done to "rotate" the time axis 90°, that is, replace t by it. This trick sometimes transforms a Feynman integral into a "Wiener" integral, familiar from the study of Brownian motion. Wiener integrals, in turn, can be solved. The solution is then rotated back 90° to get the solution of the quantum mechanical problem.
12 Manin (1989).
13 I used the term "formal analogy" somewhat differently in "The application of mathematics in natural science" (1989).
14 How Euler avoided errors, lacking a deductive foundation for analysis, remains a mystery. Even if he relied on unformulated mathematical principles, we do not know to what extent these might coincide with the actual subsequent developments initiated by Cauchy and Riemann. A careful study of Euler's methods might yield results of great interest to historians, philosophers, and mathematicians. For these points I am indebted to Larry Zalcman.
15 That is, a function of a complex variable is analytic in a region of the complex plane iff the region is a union of open disks, and on each of these disks the function is given by (the same or different) convergent power series. Note that every power series of a complex variable converges nowhere, everywhere, or in an open disk of some radius of convergence R – in the complex plane.
16 Not quite, since $\Gamma(n + 1)$ turns out to be $n!$
17 The factorial function appears ubiquitously in the integrals used in physics. The function $f(n)$ is given by a series containing infinitely many definite integrals, so we shall have infinitely many applications of the Γ function.

18 What is meant is that we get the right numbers out of a process that, behaviorally, resembles analytic continuation. Literally, of course, the sentence is mathematical nonsense: a nonconvergent series does not define a function, and so cannot be said to be continuable.
19 For an elaboration of this point, see Steiner (1989).
20 Quantum electrodynamics is a relativistic theory where four is the relevant dimension.
21 The electromagnetic field surrounding the electron has infinitely many degrees of freedom, corresponding to infinitely many radiation frequencies, all of which are possible. Each of these frequencies must be taken account of; the contribution of each to the integral does not vanish as the frequencies soar into the ultraviolet region; hence the integrals are infinite.
22 I'm not, of course, asserting that these are the *only* assumptions that would legitimate the procedure.
23 A cutoff of a different nature follows from the known fact that quantum electrodynamics breaks down at high frequencies, in the sense that quantum electrodynamics is a limiting case of a more general theory (in fact, the "electroweak" theory). However, this was not known at the time that the procedure discussed here was first used.
24 Yuval Ne'eman informs me, however, that someone has come up with a way to make cutoffs consistent with relativity.
25 Namely, by the substitution, a theoretical, but unmeasurable, "bare" quantity like charge is replaced by an expression in terms of the charge actually observed by measuring instruments of various energies.
26 It is a tribute to the ingenuity of the human race, by the way, that we can even measure such insignificant quantities as the magnetic moment of a single electron so that the prediction can be checked.
27 Wittgenstein (1978: 122). He says "mathematicians" but what he really means is philosophically articulate mathematicians like Hilbert.
28 Diamond (1976: 209ff.).
29 It would be unjust, by the way, to attribute to Wittgenstein a complaisant attitude toward mathematical rigor or to consistency, or lack of appreciation for the work of Weierstrass and Cauchy in "cleaning up" the calculus. He reserves his ridicule for the idea that, in order to apply a mathematical formalism, it must first be protected against "hidden" contradiction by a metamathematical consistency proof. Consistency in a theory is certainly a virtue for Wittgenstein. But it is no special sort of virtue.
30 The assertion ought to be supplemented by an analysis of the notion of a "mathematical rule" which would relieve the apparent paradox that following mathematical rules can sometimes yield pseudo-mathematics. I hope to provide such an analysis in future work.
31 By this is meant the actual laws of the brain's activity, not the idealized normative laws of logic.
32 The applicability of mathematics in science is sometimes accounted for by the claim that "the world is mathematical," that is, some form of Pythagoreanism. Pythagoreanism is, I take it, irrelevant to the

present reflections, since we are referring to pseudo-mathematics here – unless, of course, the "bad" mathematics of the present prefigures the "good" mathematics of the future.
33 Shmuel Elizur, Daniel Amit, Nathan Malkin, Yuval Ne'eman, Joel Gersten, Barry Simon, Ithamar Pitowsky, Larry Zalcman, Harry Furstenburg, David Kazhdan, and Shlomo Sternberg helped me with the physics and mathematics described in this essay even when, on occasion, they disagreed with my thesis; they are not responsible for any distortions. Thanks also to Sidney Morgenbesser and Isaac Levy at Columbia, and to Eddy Zemach, Avishai Margalit, and my other colleagues in Jerusalem, for philosophical insights. My research was supported by the Israel Academy of the Arts and Sciences, for which support I am grateful.

REFERENCES

Diamond, C. (ed.) (1976) *Wittgenstein's Lectures on the Foundations of Mathematics*, Ithaca, NY: Cornell University Press.
Feynman, R.P. (1967) *The Character of Physical Law*, Cambridge, MA: MIT Press.
Gell-Mann, M. (1987) "Round table on internal symmetries," in M.G. Doncel, A. Hermann, L. Michel, and A. Pais (eds) *Symmetries in Physics (1600–1980)*, Barcelona: Servei de Publications, UAB.
Kitcher, P. (1981) "Mathematical rigor – who needs it?," *Nous* 15:469–93.
—— (1983) *The Nature of Mathematical Knowledge*, Oxford: Oxford University Press.
Lakatos, I. (1976) *Proofs and Refutations*, Cambridge: Cambridge University Press.
Manin, Y. (1989) "Strings," *The Mathematical Intelligencer* 11: 63.
Steiner, M. (1975) *Mathematical Knowledge*, Ithaca, NY: Cornell University Press.
—— (1989) "The application of mathematics in natural science," *Journal of Philosophy* 86: 449–80.
—— (1983) "The philosophy of mathematics of Imre Lakatos," *Journal of Philosophy* 80: 502–21.
Van Fraassen, B.C. (1980) *The Scientific Image*, New York: Oxford University Press.
Wittgenstein, L. (1978) *Remarks on the Foundations of Mathematics*, Cambridge, MA: MIT Press.

7

FOUNDATIONALISM AND FOUNDATIONS OF MATHEMATICS*

Stewart Shapiro

SUMMARY

This essay is a study of foundationalism and rationalism in mathematics, and the relationship between one's views on foundationalism and the appropriate or best, logic. I suggest that foundationalism or rationalism explicitly or implicitly dominated work in logic and foundations of mathematics until recently, most notably logicism and the Hilbert Program. But foundationalism has now fallen into disrepute. It might be recalled that it does suggest a rather straightforward criterion for evaluating proposed logics and foundations: namely, self-evidence. The issues here concern how much of the perspective is still plausible and how logic and foundations are to be understood in the prevailing anti-foundationalist spirit. How are candidate logics to be evaluated now? One common orientation seems to be to regard logic and, perhaps, mathematics in general as an exception to the prevailing anti-foundationalism, or anti-rationalism – a sort of last outpost as it were. Against this, I argue that we have learned to live with uncertainty in virtually every special subject, and we can live with uncertainty in logic and foundations of mathematics as well. In like manner, we can live without completeness in logic, and live well.

The names of once exotic, but now commonplace, mathematical entities indicate a preference among mathematicians for security. Beyond the "natural" numbers, there are "negative," "irrational," "transcendental," and finally "imaginary" numbers. Fortunately, the profession has always had its bold, imaginative souls, but it seems that they do not get to provide the official names.

As a first approximation, define *foundationalism* to be the view that it is possible and desirable to place a given branch of mathematics on a completely secure foundation. For better or worse, foundationalist programs explicitly or implicitly dominated much of the work in logic and foundations of mathematics until recently, but, for good reason, foundationalism has now fallen into disrepute. It might be recalled that foundationalism does suggest a prima facie straightforward criterion for evaluating proposed logics and foundations, self-evidence. It seems that, in retrospect, the concept of self-evidence has proved elusive.

In the preface of his landmark work on the foundations of analysis, *The Continuum* (1987), Hermann Weyl wrote:

> With this essay, we do not intend to erect ... a ... wooden frame around the solid rock upon which rests the building of analysis, for the purpose of persuading the reader – and ultimately ourselves – that we have thus laid the true foundation. Here we claim instead that an essential part of this structure is built on sand. I believe I can replace this shifting ground with a trustworthy foundation; this will not support everything we now hold secure; I will sacrifice the rest as I see no other solution.

In this work, Weyl proposed predicative restrictions on analysis, and later he adopted intuitionism.

My concern in this essay is to examine how much of the perspective of foundationalism is still plausible, and how logic and foundational studies are to be understood in the prevailing anti-foundationalist spirit. How are candidates to be evaluated now? It seems that philosophical movements spawn tendencies that remain long after the views themselves are dismissed, at least publicly.

The short answer is that just as we have learned to live with uncertainty in virtually every special subject, we can live with uncertainty in logic and foundations of mathematics, and we can live well. I accept and support the first of Weyl's claims: classical mathematics does not have an absolutely secure foundation. It is truly a house built on sand. But I reject the view that there is something wrong with this. Mathematics does not stand in need of a secure basis; the building does not fall down when we realize its lack of a grounding in bedrock. There are important roles for foundations of mathematics, but providing absolute security is not among them. Weyl's own metaphor is suggestive: the "foundation" is wood; the main structure, mathematics, is rock.

As I see it, the legitimate purposes of foundational studies for a branch of mathematics include the codification of its reasoning, forging connections with other areas of study, and exploring its conceptual and semantic presuppositions. None of these needs be construed as yielding absolute security, a foundation in bedrock, nor should a foundation be rejected for failing to provide this. If there are doubts about the branch in question, foundations might provide *some* security. Perhaps reservations concerning negative or rational numbers were alleviated when it was shown that they can be modeled in, or reduced to, the natural numbers. Similarly, real numbers can be thought of as sets or sequences of rational numbers, and complex numbers can be pairs of real numbers. At best, however, the security is relative. One who is content with the arithmetic of natural numbers has no reason to reject rational numbers.[1] In any case, quieting skepticism is, at best, a small part of the enterprise. There is surely a gain in insight when whole branches of mathematics are codified, and relations between them are articulated. Such studies have opened fruitful lines of research in mathematics itself. Entire fields, like axiomatic set theory and recursive function theory, have been spawned. Foundations also has philosophical value. The connections between fields often allow a reduction of ontology. Nowadays, it seems, sets can be the only mathematical items in one's ontology. More substantially, if one adopts a philosophical view that places limits on the conceptual resources available to humans, as knowers, one would like to ascertain how much mathematics is available. A coherent grasp of the semantics of ordinary mathematics, and its conceptual presuppositions, is surely a helpful, if not necessary, ingredient in this assessment.

I. VARIATIONS AND METAPHORS

Let P be a body of knowledge, or a field of study. The main example, of course, is mathematics, or one of its branches. Perhaps logic can also be included, but that will be considered later. A foundationalist program for a natural or social science is not worth considering, at least not now.

Define *strong-foundationalism* for P to be the view that *there is* a single foundation for P, one that is absolutely secure or, failing that, as secure as humanly possible. The metaphor is that in providing a candidate-foundation one is attempting to "dig" one's way to *the* ultimate basis for belief in the theses of P, an attempt to expose what is already there, for all to see.

Define *moderate-foundationalism* for P to be the view that it is possible to provide *at least* one foundation for P, again either absolutely secure or as good as possible. Here, the metaphor is that in putting forward a candidate, one is not necessarily "digging down," but rather is "building up" from the "bedrock" to P. Instead of exposing what is already there, one is providing something new in the development of P. A moderate-foundationalist program is, in effect, a *re*construction of P. On this view, the question of whether there may be more than one foundation is put aside. Alternatives are not necessarily competitors. For all we know, the field can be rebuilt in more than one way.

The difference between these views is not that important here, and I do not wish to engage in exegetical issues concerning the sort of foundationalism assumed by various authors. So define P-*foundationalism* (simpliciter) to be the disjunction of the two views. In either case, if the "bedrock," the material *of which* foundations are made, is self-evident, then well and good. If not, the foundationalist can be consoled with the belief that he has reached a (or the) place beyond which no further building, rebuilding, digging, or general "securing" is humanly possible. To continue the metaphors, even if it is possible to dig below the bedrock, there is no point.

The name "foundationalism" and the above metaphors suggest an analogy with the construction industry. The analog of the building is the field of study P. One striking *disanalogy* is that historically, at least, P was not developed like a building by *first* laying a foundation and then adding floors, walls, ceiling, more stories, etc. In mathematics, foundational activity takes place either (long) after the field P is established or at most, is contemporaneous with the development of P. More importantly, the foundation of P has at least some independence from P itself, more so than the foundation of a building has from the building.

The last hundred years (or so) has seen a number of major foundational efforts. I consider two of them, logicism and the Hilbert Program.[2] Both have been evaluated and analyzed in enormous detail, but I will indulge in a few words here, since the issues surrounding the present project developed in this context.

In both cases, the plan of reconstruction has something in common with the more grandiose project adopted by Descartes. It is the familiar axiomatic method, with attention to foundationalist detail and logic. A core basis of the tenets of the field are identified and labeled "axioms" or "definitions." Then one shows how the entire body of knowledge can be developed step-by-step from the axioms by methods of reasoning. Unlike prior axiomatic treatments (such as Euclid's *Elements*), it is important for both foundationalist programs that the *logic* itself be made explicit. One must precisely delimit the notion of *correct inference*. That is, both the basis and the methods of proof are to be fully stated, without gaps.

Probably the most important difference between the two programs concerns how the security is to be guaranteed. In the Hilbert program, one is to present the logic as a *calculus* – as a collection of mechanical rules operating on the formulas of a formal language. Reliability is secured with a *proof* that the system so construed is "consistent." This, of course, requires a meta-theory in which to carry out the consistency proof. The subject-matter of this meta-theory is the object language formulas, regarded as strings on a finite alphabet, strings that might as well be meaningless. The notion of a formal deductive system as calculus is apt. Indeed, the program has been called "logic as calculus." Noting that the structure of strings is equivalent to that of natural numbers, the meta-theory is to be finitary arithmetic, regarded by all as unquestioned. As pointed out by Tait (1981), this may not produce *absolute* reliability, in that the meta-theory does not get an independent foundation, but it is the best we can do. There is no *more* preferred or more secure standpoint from which to criticize the envisioned meta-theory used to secure the reliability of the object language theory.

With logicism, the older program, the situation is not like this. The theory/meta-theory distinction is rejected, out of hand. As emphasized by van Heijenoort (1967) and Goldfarb (1979), the envisioned language and deductive system are to be characterized with sufficient precision to allow them to be used mechanically, but this is not the crucial aspect of the program. The language is to be completely interpreted, the diametric opposite of the idea that it be considered a meaningless formal system. Moreover, the language should be universal in the sense that it applies to any subject-matter. As Frege (following Leibniz) elegantly put it, the language is to be at once a *calculus ratiocinator* and a *lingua characteristica*. It captures the ideal of complete justification. There can be no other standpoint from which to view and study the language of logicism – no meta-theory. It is already universal, indeed *the* universal language. Thus, the axioms and inferences of the system must *themselves* be self-evidently reliable. There is nothing extra we can do to "establish" reliability in logic. An ordinary-language inference is reliable just in case it corresponds to an inference in the universal language. The perspective has been called "logic as language."[3]

As Goldfarb (1988) notes, this program invites a charge of circularity. If we do not accept the proposed language and deductive system as *characterizing* reliability, then we can doubt the reliability of the system. To follow the construction metaphor, the system is to provide a blueprint for the building after (or while) it is built. It is to assure us that it will not collapse. If we doubt that the given system is the "blueprint," we will not be assured of anything. But, as Goldfarb adds, the circle is hard to resist, especially in retrospect. The logicists claim that they are only making explicit the standards of rigor already implicit in at least careful mathematical practice.

Both programs require that the language (or languages) in question be formulated with sufficient precision to be used mechanically. The Hilbert Program, logic-as-calculus, allows another perspective not available to logicism. Since the languages of the Hilbert Program are to be regarded as uninterpreted, one can discuss *alternative interpretations* of their (non-logical) terminology. That is, the logic-as-calculus perspective allows the development of a model-theoretic semantics. Indeed, formal logic spawned a model theory as well as a proof theory. This dual aspect of formal languages plays a large role in the present study. With logicism, however, the language is already completely interpreted, and thus alternate interpretations are not countenanced. In effect, the languages of logicism do not have *any* nonlogical terminology.[4]

Both of these programs failed to achieve the foundationalist goal and, for various reasons, few people seriously hold out hope for repairs.[5] Briefly, logicism failed because it was concluded that the reduction of mathematics to logic requires the incorporation of set theory into the universal language and logic. The development of this was rather arduous – the proposed blueprint proved hard to read. There is no question concerning the power and fruitfulness of axiomatic set theory, the offspring of this work, but history shows that it is anything but self-evident. As for the Hilbert Program, with hindsight it is surely reasonable to insist that any deductive system within its framework must be effective[6] and, of course, consistent and sufficient for elementary arithmetic. Gödel's Second Incompleteness Theorem shows that one cannot prove the consistency of such a deductive system D in D, much less in a theory weaker than D, one *more* self-evident.

This is not to belittle the enormous accomplishments of these programs. They spawned entire branches of mathematics, and, after all, set theory *is* a foundation of mathematics (as is category theory). But it does not satisfy foundational*ism.*

Writing after the dust had settled, Skolem (1952) sketched three possible desiderata of foundational research in mathematics. The "logicist" desire is

> to obtain a way of reasoning which is logically correct so that it is clear and certain in advance that contradictions will never occur, and what we prove are truths in some sense.

Another, more modest aim is only "to have a foundation which makes it possible to develop present day mathematics, and which is consistent so far as is known yet." This is called the "opportunistic" outlook. We might say that it is foundations without foundationalism. Skolem states that it has the "unpleasant" feature that we are never certain when the foundational work is complete. He makes liberal use of the construction metaphor:

> We are not only adding new floors at the top of our building, but from time to time it may be necessary to make changes at the basis.

It can be agreed (I presume) that this is not a sound way to erect buildings, but part of the present goal is to question the metaphor. Skolem's third "desire" is the "Hilbert Program," which is characterized as the result of "giving up the logicist standpoint and not being content with the opportunistic one."

Skolem was, of course, well aware of the difficulties with these foundationalist programs, but, apparently, he was reluctant to give up on foundationalism altogether. Today, some decades later, it is generally held that *nothing* can satisfy the requirements of foundationalism. That level of security cannot be attained. We are left with Skolem's "opportunistic" option, and anti-foundationalism.

This raises two kinds of questions. First, *to what extent* is foundationalism impossible to sustain? Perhaps *part* of the goal can be achieved. Second, what are the purposes and goals of an "opportunistic" foundation? Why do we want one, and how do we evaluate candidates? These questions are, of course, interrelated.

II. FOUNDATIONS AND PSYCHOLOGISM

The metaphorical talk of foundational studies, building and digging, suggests that it may be a psychological enterprise – even a therapeutic or clinical one. Returning to our generic field of study P, the appearance is that one sees some value in P and either feels threatened (e.g. by antinomies) or worries about future attacks on P. The foundation – strong or moderate – seems designed to alleviate these worries and make the P-theorist feel secure once again.

There is, I believe, some historical support for this orientation. Recall, for example, Hilbert's (1925) quip about never having to leave Cantor's paradise. In many minds, however, this outlook is dangerously misleading. Foundations is an *epistemic* endeavor, not a psychological one. It is conceptual, not empirical; objective, not subjective. The aim of foundations is to clarify or develop a basis for *knowledge* in the theses of P. From this perspective, the question is not "What can make us feel better about believing, pursuing, or practicing P?," but rather "Why or how do we know P?" If we *only* want to relieve our anxiety about P, then we could engage a (real) therapist (or a hypnotist, or a surgeon, or we could ingest certain drugs). There is surely more to the goal of foundational studies than a license to go on working in P with reckless abandon.

In short, we want an epistemic *justification* of P, not a *feeling* of certainty. A foundation should conform to objective *norms* of justification and knowledge. Foundationalism, so construed, is an anti-psychologism. But, of course, the central matter at hand is the question of what counts as a justification, and there is the rub.

This charge against the role of psychology in foundational work is, I believe, unfair. The P-theorist certainly does not want to *merely* go on doing P, and certainly not with abandon. The antinomies and threats do, after all, tell us something, and we have to figure out what precisely the potential problems are, and how to avoid them. Both advocates and opponents of psychologism agree that P plays a certain role in our overall intellectual life, and we need some assurance that this role can continue. So, on all accounts, the desire is to put P on solid ground with reason. It would be a distraction to argue for this in detail, but I would suggest that, in some cases at least, the attacks on P leave an unclarity concerning just what P is.[7] It is not all that clear *how to go on* with the practice of P. At least part of the goal of foundations is to provide guidance on this. This goal, at least, is independent of the issues of psychologism. Indeed, it is independent of foundational*ism*.

I do not wish to aid a revival of an extreme version of psychologism, the idea that there is nothing more to epistemology than psychology. But I also reject the other extreme, that psychology is in no way relevant to epistemology. Indeed, even if we did have a candidate for a foundationalist system, the relevant epistemic issues would be difficult, perhaps impossible, to resolve on neutral ground. Poincaré launched a broad attack on the emerging field of logic, partly on psychologistic grounds. He argued that the new logic is certainly not self-evident, and that the translations of simple arithmetic statements into its language are far *less* evident than the original statements. The supposed foundation is less solid than the enterprise it is supposed to support. Russell replied that this charge confuses psychological certainty with epistemic warrant. In a rejoinder (1909: 482), Poincaré held his ground:[8]

> M. Russell will doubtless tell me that these are not matters of psychology, but of logic and epistemology. I shall be driven to respond that there is no logic and epistemology independent of psychology. This profession of faith will probably close the discussion.

Let us pursue this for a moment.

The main question here is whether epistemic matters *can* ultimately be divorced from psychological ones. We need an account of *what it is* for an argument to be an epistemic warrant, quite independent of whether most thoughtful people (in the right frame of mind) will, as a matter of fact, be convinced by it.

If logicism could get off the ground, at least in part, it would provide the beginning of a response to this query. An argument is a true epistemic warrant if it conforms to the ideal of reason as codified by the system. As noted above, this would be a circle, but perhaps it is not all that vicious. It would depend on how plausible the logicist system is, qua logicist system.

Notice that this is a contrary-to-fact thought-experiment. Logicism *did* fail and we still have no acclaimed account of what an epistemic warrant is. Whether we can obtain one may depend on the extent to which *part* of the logicist program can be salvaged. I will briefly return to this in the next section. But, for now, let us continue the thought-experiment. Suppose that we did have a candidate (partial) logicist system that is purported to capture the notion of epistemic warrant. We would thus be confronted with the "circle" and would have to decide whether to "enter" it. It is surely out of the question to request the proponent to *prove* to us that the system does capture the notion of epistemic warrant – or even that the system is consistent. Again, the very matter at issue is what *counts* as a proof. Recall the logicists' rejection of the value of meta-theory.

Continuing the subjunctive mood, there would be two questions to be settled about the candidate deductive system. First, are the inferences it sanctions in fact correct and, second, is the system exhaustive – does it sanction *every* correct inference? These questions represent a sort of soundness and completeness. There are two philosophical stances that would be available to our logicist (plus the possibility of a combination). One is to make a concession to psychologism and assert that the sanctioned inferences all *seem* certain and, as far as we can tell, every inference we are in fact certain of (on reflection) is codified by the system. Of course, this *is* a concession, but it does not suffer from the worst psychologistic excesses. The concession is limited to the initial justification of the system. The other stance is to postulate a faculty of normative *epistemic* intuition, a faculty outside the purview of psychology.[9] The claim is that we can recognize at least *some* correct inferences and that we do in fact recognize the inferences sanctioned by the proposed deductive system. However, I do not see how either the concession or the intuitive epistemic faculty could assure us that the system is *exhaustive*, that *every* correct inference is sanctioned thereby. It is surely far-fetched to hold that a faculty of epistemic intuition could tell us this. At best, it would judge individual inferences. Perhaps our logicist can live with the possibility that the system may have to be expanded one day. This is an "incompleteness" of sorts.

Notice that on both of these stances the program is a familiar one in philosophical theory-building. We start with our "intuitive" beliefs, instincts, dispositions to judge, etc. Formalization begins with these as "data." From that point, the theory can become a guide to intuition, as intuition is a guide to theory. The difference between the two stances is over the nature of the *original* "intuition" – whether it is psychological or epistemic, natural or normative, etc. (see Resnik 1985).

At the risk of ridicule, I confess to some sympathy for the views of both Poincaré and Russell. I agree with Poincaré on the original motivation or origin of logical theory and foundational study. That is, at the outset of theory, when considering the "data" with which one is to begin, Poincaré is correct that "there is no logic and epistemology independent of psychology." It is, in effect, a working hypothesis that the inferences and propositions that seem correct are correct. The stronger our feeling of certainty about a certain item, the less likely it will be challenged or revised in light of theory. In some cases, we simply cannot imagine revision, at least not now. This applies both to logic, with our intuitions concerning correct inference, and to special fields like arithmetic, with our intuitive feelings towards, say, simple sums. The products of the translations provided by the logicists are certainly less evident than the simple identities they translate. However, once theory has begun, and shows signs of success (whatever that may be), then our intuitions can be guided by its light. Indeed, at any given time, our intuitions are the product of our background and training, and that is, at least in part, the product of previous intellectual endeavors, more theory.

An interesting case study is Bolzano's (1817) proof of the intermediate value theorem. It was based on careful, rigorous definitions of continuity, limit, and the like, and it included a demonstration of the Bolzano–Weierstrass theorem, that every bounded, infinite set of points has a cluster point. Bolzano's proof was certainly a major achievement, a hallmark of modern analysis. A century or so earlier, however, mathematicians would have scoffed at the idea that the intermediate value theorem *needed* a proof. What could be more evident, indeed more self-evident, than the statement that a continuous curve which passes from below the x axis to above it must cross the axis somewhere. This is what it *means* to be a continuous curve, or so people thought.[10] At some point, however, it became clear that intuitive conceptions of, say, continuity, were not sufficient. To use a Kuhnian phrase, the paradigm was breaking down, and definitions were needed (see Coffa (1982) and the last chapter of Kitcher (1983)). It became intuitively evident that what was previously (thought to be) self-evident was no longer evident.

On the other hand, once a successful theory, or meta-theory, is in place, there is a coherent distinction between psychology and logic, and between psychology and epistemology. That is what the theory provides. Even in mature foundational studies, however, it is a truism that one cannot continue to formulate formal meta-meta-...-theories. At some point (usually rather quickly) we reach an informal language, a language to be *used*, not studied, and used without the benefit of more theory.[11] At this level, again, Poincaré is correct that "there is no logic and epistemology independent of psychology."

As for the details of the debate, however, I think that the surviving variants of the logicistic systems are more or less correct as far as they go – against Poincaré: that is, correct to the extent that they are consistent (Frege, and Russell's paradox) and not overly cumbersome and unintuitive (Russell and Whitehead, and ramified type theory and the translations of arithmetic identities). But (unlike Poincaré) I have the benefit of historical hindsight. The systems have proven fruitful and they do conform to our *current* intuitions concerning correct inference. Theory has guided intuition just as intuition has guided theory.

Without question, the most widely acclaimed surviving fragment of the foundationalist programs is classical first-order logic. It is a central component of epistemology, *as opposed to* psychological certainty. But, given the failure of the foundationalist programs, we should not be so smug as to hold that first-order logic is all there is. Indeed, is any given formal system, no matter how well established (whatever that might mean), the *whole story?* Does it capture *every* correct inference? In the contemporary scene, each candidate logic is cast in a formal object language, or a class of such languages. There are two questions to be asked of each logic. First, does it codify every correct inference *in that language* (or in those languages)? I argue below that there is reason to think so in the case of classical, first-order logic. Second, is the language itself (or the type of language) sufficient to capture every inference of mathematics (or whatever is being studied) in a perspicuous manner? It is hard to see how one could establish an affirmative answer to this for any formal language. I have argued elsewhere (1985 and 1991; see also Boolos 1975) that first-order languages are *not* adequate. We should further weaken foundationalist aims and accept, at least provisionally, a stronger logic, one less secure and less "certain" than first-order.

III. TWO CONCEPTIONS OF LOGIC

There are, I believe, two different orientations toward logic, even today. Distinguishing these can help expose philosophical prejudices and can forestall discussion at cross-purposes. I call the orientations the *rationalist* and the *semantic* conceptions of logic. They have different criteria and suggest different kinds of systems. As will be shown, there is a sense in which the criteria coincide when attention is restricted to first-order systems. First-order logic is, after all, complete. But higher-order systems are not. The conceptions are not necessarily competitors, but (to undermine the dialectical drama), I think the semantic one is more plausible. The two conceptions are not exhaustive of current conceptions of logic.

Rationalism

In both foundationalist programs, the axiomatic method has two stages (unlike, say, Euclid's *Elements*). One gives axioms *and* one gives rules for deducing theorems – a deductive system. In principle, of course, there is not much difference between axioms and rules. Any axiom can be thought of as a rule of inference, and many rules can be recast as single statements, in the form of conditionals. In practice, however, axioms often concern the special subject under study, while rules of inference usually concern the logic. The latter are rules for deriving conclusions, valid in any field.[12] This is the important distinction here. Perhaps the concession to anti-foundationalism can be limited to the axiom-providing stage, to the search for self-evidence among the truths of each branch. On this view, for a given field P it may not be possible to locate a sufficient core of self-evidence among its truths, but it *is* possible to provide rules of inference that are self-evidently correct – or self-evidently as reliable as possible. We cannot be absolutely sure that our starting axioms are true, but *if* they are, *then* we can be certain that the deductive system will not lead us astray.

This retrenched program is an attempt to salvage "logic" from logicism. Following Wagner's (1987) lucid account, a logical system is seen as the ideal of justification, at least *relative justification*. We can be sure that we will not deduce anything false from true axioms. Wagner calls this the *rationalist conception of logic*.

As might be surmised, I do not believe that even this much of the logicist program can be salvaged. My own anti-foundationalism is more thorough. Here, I sketch what logic would be like on the rationalist account, followed by what I take to be an alternative conception of logic, a semantic one.

The central concern with rationalistic logic is, of course, proof. So its main item is a deductive system. It should codify all and only the inferences that conform to the postulated standard of (relative) ideal justification. Thus, as in the previous section, there are two sorts of questions to be answered of each candidate. First, are all the sanctioned inferences actually instances of ideal justification? Call this the issue of *conformity*. Second, are all ideally justified inferences thereby sanctioned. Call this the issue of *sufficiency*. Conformity and sufficiency are, in a sense, informal counterparts of soundness and completeness. They concern the match (in extension) between a precisely defined formal notion and an intuitive, pre-formal counterpart.[13]

Clearly, it would not advance knowledge or understanding to *define* the deductive system as follows: a pair $\langle \Gamma, \Phi \rangle$ is a correct inference if and only if Γ is sufficient to justify Φ. It must be described in some other way, and thus the problem of conformity arises. Are the codified inferences in fact correct? Since rationalism is a limited version of logicism, we may expect a similar disavowal of a foundational role for semantics and meta-theory. As will be seen, this would be premature, but it is certainly correct that meta-theory cannot help with the *conformity* issue. The deductive system is supposed to constitute ideal justification. There is no more secure perspective from which to justify this. No proof can have premises more evident than the system it is meant to secure. In particular, a proof of consistency would not help. Thus, the rationalist sides with Frege against Hilbert.

It seems plausible that a rationalist deductive system ought to be at least a crude model of the *process* of justification. It might be constructed by specifying a collection of *immediate inferences* and defining a pair $\langle \Gamma, \Phi \rangle$ to be correct, sanctioned by the deductive system, and only if there is a finite sequence $\Phi_1 \ldots \Phi_n$ of formulas in the given language such that Φ_n is Φ and each Φ_i is either in Γ or follows from previous formulas by an immediate inference.

On this model, the conformity of a deductive system would come down to whether each immediate inference is, in fact, self-evidently correct (or at least correct).[14] It seems reasonable to require of any deductive system (that purports to conform to rationalism) that this be true and, moreover, that it be knowable beyond dispute. We should be able intuitively to check each immediate inference, or each immediate inference scheme, for correctness. That is, it must be assumed that our intuitive abilities go that far.

It is too much to expect our powers of recognition to determine whether a given deductive system contains *every* justifiable inference, however, even for a fixed language L. That would require an intuitive ability that not only checks individual inferences for (self-evident) correctness, but also provides some sort of pre-theoretic "map of logical space" (so to speak). Intuition would have to determine its boundaries. This, I believe, is more than a rationalist can reasonably postulate, much less defend. Thus, for the question of sufficiency, the standards must be relaxed. Self-evidence is out of the question. I suggest, then, that sufficiency is a *theoretical* question, not an intuitive one, and thus seems to require some theory, or meta-theory, for an answer – this despite the logicists' rejection of meta-theory.

When considering the rationalist conception of logic alone, this is the end of the line (if I have not gone too far already). To continue, something more must be said about the nature of ideal justification. So far, it has been taken to be more or less primitive, or "intuitive." The only thing said is that justifiability is related to "self-evident correctness" (via finite sequences of formulas). Here, perhaps, we have a role for semantics, and this leads to the other conception of logic.

The semantic conception

Wagner (1987) is correct that the rationalist conception of logic underlies the work of the logicists, most notably Frege. I might add that it is a natural successor (or correction) to earlier ideas of logic as the "laws of thought."[15] As Wagner concedes, there is a second conception of logic, a semantic or model-theoretic one. I will not venture any detailed statements about its pedigree, but I believe it can be traced to Aristotle. A modern exponent is Tarski (see, for example 1935).

In textbooks and courses in elementary logic, validity is often characterized informally in either semantic or modal terms, at least at first. One asserts that an argument is "valid" if it is not *possible* for the premises to be true and the conclusion false, or if any *inter-*

pretation that makes the premises true also makes the conclusion true. The semantic conception of logic is a natural explication of these definitions. As above, one provides a collection of models or interpretations, and characterizes "correct inference" in terms of (semantic) validity. That is, $\langle \Gamma, \Phi \rangle$ is valid if and only if Φ is true in every model in which every member of Γ is true. It is this notion that is regarded as primary.

The semantic notion of logic depends on a distinction between the "logical terminology" and the "nonlogical terminology" of the language in question.[16] The nonlogical items are those whose interpretations or referents vary from model to model, while the logical items are those whose meaning is the same in all the models. The interpretation of the logical terminology is provided by the model-theoretic semantics as a whole.

The extension of semantic validity depends on where the line between logical and nonlogical is drawn. Focus, for the moment, on a common (first-order) language of arithmetic, augmented with a predicate letter D. Let s be the name of the successor function, and consider the following two inferences:

The ω-rule: From $\{D(0), D(s0), D(ss0), \ldots\}$ infer $\forall x D(x)$.
From $\{D(0), \forall x(D(x) \rightarrow D(sx))\}$ infer $\forall x D(x)$.

In standard first-order semantics, neither of these is valid. Consider an interpretation whose domain is the set of all countable ordinals, with the numerals and successor function given their usual (von Neumann) readings, and let the extension of D be the set of finite ordinals. On this interpretation, all the premises of these inferences are true and the conclusions are both false.

The informal "reason" one might give for the invalidity of these inferences is that the premises *only* say that D holds of 0, $s0$, $ss0$, etc.; they do not say that 0, $s0$, $ss0$, etc. are *all* the numbers there are. The above "counter-interpretation" exploits this by *re*interpreting the numerals as ordinals. Such is common practice. But suppose someone were to challenge this and claim that both inferences are valid *as they stand*. The claim is that the proposed counter-interpretation is not a legitimate interpretation of the language. Indeed, the numerals and successor symbol are logical terms, not subject to reinterpretation in the models of a semantics. Even if we could, we do not have to explicitly state in the premises of an argument that 0, $s0$, $ss0$, etc. are all the natural numbers there are.[17]

Notice the analogy between this disputant's remarks and ours if we were confronted with a claim that

from P&Q infer P

is not valid, due to the interpretation (over the natural numbers) of P as $s0 = 0$, Q as $s0 = s0$, and the ampersand as "or". We would (justly) protest that this is not a legitimate interpretation. One does not have to state *in* the argument that "&" means "and" (even if we could).

In effect, if there is to be *no* nonlogical terminology (the opposite of logicism) then no inferences are semantically valid. This is surely an extreme not worth pursuing. In my view, however, the distinction between logical and nonlogical terminology is not *a priori*. It is a theoretical matter, adjudicated on holistic grounds, on how useful and illuminating the resulting model theory is.

On the semantic conception of logic, the plausibility of a candidate semantics depends on the extent to which the class of models corresponds to the intuitive notion of "interpretation" or "possible world," the one active in pre-formal judgments of validity. Analogous to the rationalist program, there are two questions for each candidate. First, does the formal semantics *conform?* That is, is each model thereof a legitimate interpretation of the given language? Second, is the candidate semantics *sufficient* in the sense that it has "enough" models? In other words, is it the case that, for each set Γ of formulas of the language, if there is an "intuitive" interpretation that makes every member of Γ true, then there is a model in the semantics that makes every member of Γ true?[18]

To be sure, our "intuitions" concerning correct inference are relevant in answering these questions in a given case, but from this perspective, they are not the whole story. Here it is much clearer that our pre-formal judgments are tentative data, subject to correction. Suppose, for example, that there were an inference that is "intuitively correct" but not semantically valid (according to a candidate model theory). That is, suppose that the inference is obtained by a chain of short steps, each of which is "self-evidently correct" (or very certain), and suppose that in the candidate semantics there is an interpretation that makes the premises true and the conclusion false. There would be tension; *something* must be changed. One option would be to modify the semantics and exorcise the offending "interpretation" (preferably in a principled way, not *ad hoc*). For example, the boundary between the logical and the nonlogical terminology might be moved. Or we could modify our "intuitive" judgments about the inference in question. Of course, in such a case, *some* explanation of where our preformal intuition went wrong is called for. For example, we could locate an offending step in the argument chain and make explicit an "assumption" (or, to use Lakatos's phrase, a "hidden lemma") of the derivation. On the other hand, if the semantics is sufficiently entrenched, the "explanation" might consist of little more than the *observation* that the offending argument is not (semantically) valid.

I might add that, unlike the rationalist conception, judgments about correct inference are not the only major items involved in evaluating candidate model theories. The language in question is used to *describe* structures as well as to justify propositions, and we have at least tentative, pre-formal beliefs or, for lack of a better term, "intuitions" about possibility and interpretation. After all, the above *informal* semantic-modal characterization of validity is phrased in those terms, and when newcomers to logic hear it, they have *some* idea of what we mean. To be sure, such pre-formal ideas, and original intuitions, must be refined by theory, radically in some cases. I conclude, for now, that evaluating semantics is a complex, holistic matter. It is not a foundationalist program, although it might augment one. Let us turn to the relationship between the two conceptions of logic.

IV. MARRIAGE: CAN THERE BE HARMONY?

If we focus on a *fixed* language L, it is possible that both conceptions of logic converge on a single candidate (provided, of course, that one regards both conceptions as legitimate). It might be the case that an inference conforms to ideal justification if and only if it is semantically valid. I would suggest that this in fact occurs in the case of first-order languages, but let us be more abstract for the moment.

Suppose that one develops a deductive system as a candidate for the codification of ideal justification, and develops a formal semantics as a candidate for the pre-formal semantic conception of logic. Then there would be two formal structures, and relations between them could be investigated mathematically. In particular, we can try to prove soundness and completeness theorems. Soundness is the statement that every inference sanctioned by the deductive system is valid in the model theory. Completeness is the

FOUNDATIONALISM AND MATHEMATICS

converse, that every inference valid in the model theory is sanctioned by the deductive system. The question at hand is whether such results would strengthen *either* the claim that the deductive system is correct for ideal justification *or* the claim that the semantics is true to the pre-formal notion of semantic validity. This, it seems, depends on the relationship between the pre-formal notions themselves.

In the imagined scenario, there would be several "philosophical" questions. We have encountered some already. Let me recapitulate, and add to the list.[19]

(1a) The proposed deductive system is a correct codification of ideal justification for the language at hand. That is:
  (1a1) If the deductive system contains the inference $\langle \Gamma, \Phi \rangle$, then Φ can be justified on the basis of Γ (conformity).
  (1a2) If Φ can be justified on the basis of Γ, then the deductive system contains the inference $\langle \Gamma, \Phi \rangle$ (sufficiency).
(1b) The proposed semantics is true to the pre-formal notions of "interpretation" and "validity." That is:
  (1b1) Each model in the semantics is a legitimate interpretation of the language. That is, if Φ is true in each interpretation of the language in which Γ is true,[20] then Φ is true in each model of the semantics in which Γ is true (conformity).
  (1b2) There are "enough" models in the semantics. That is, if Φ is true in each model of the semantics in which Γ is true, then Φ is true in each interpretation of the language in which Γ is true (sufficiency).
(2a) If Φ can be ideally justified on the basis of Γ, then Φ is true under every interpretation of the language in which Γ is true.
(2b) If Φ is true under every interpretation of the language in which Γ is true, then Φ can be ideally justified on the basis of Γ.

In a sense, the conformity of the deductive system (1a1) is "internal" to the rationalistic conception. I suggest above that it can probably be verified by checking each axiom and rule of inference against our intuitions concerning relative justification. On the other hand, (1a2) is more "theoretical," and we are presently looking to semantics for help. Similarly, (1b1) is more or less "internal" to the semantic conception, and we can probably rely on pre-formal intuition, at least at first. However, (1b2) is more theoretical, and requires some analysis of the purpose of the language and logic.

Item (2a) is a deeply held thesis among those who employ *both* a rationalist and a semantic conception of logic. The two *must* be related this way. Any purported counterexample would have to be eliminated, or else our intuitions would have to be repaired in its light. The intelligibility of (2a) would be challenged by someone who uses the rationalist conception alone, such as a logicist, and by someone who rejects the rationalist conception (in favor of the semantic). But if both conceptions are accepted, then (2a) is a substantive, if uncontroversial, thesis.

Item (2b) asserts that idealized human ability suffices to justify *every* (intuitively) semantically correct argument in the language. This is prima facie unlikely. Indeed, the variable in the definition of "semantically correct" ranges over all interpretations, some of which are (uncountably) infinite, while our abilities, even idealized, surely have some limit.

Notice that *if* we assume (1a) and (1b) *and* have established the soundness and completeness of the formal semantics and the deductive system for each other, then (2a) and (2b) follow. In particular, it is straightforward that (2a) follows from (1a2), soundness, and (1b2); and (2b) follows from (1b1), completeness, and (1a1). These are simple applications of the propositional calculus (hypothetical syllogisms).

But (2a) and (2b) are not the primary matters at hand. The success of the programs associated with our two conceptions of logic, when considered separately, are summed up in (1a) and (1b); and we want to examine the extent to which soundness and completeness bear on *these* questions.

For what it is worth, (1a1) follows from soundness, (1b2), and (2b). More substantially, (1a2) follows from (2a), (1b1), and completeness. Combined, (1a) follows from (1b), (2a), (2b), soundness, and completeness. Similarly, (1b1) follows from (2b), (1a2), and soundness; and (1b2) follows from completeness, (1a1), and (2a). Combined, (1b) follows from (1a), (2a), (2b), soundness, and completeness.

It can be concluded on the basis of all this that if soundness and completeness have been established, then (1a), (1b), (2a), and (2b) give mutual support to each other. Any one of these follows from the other three.[21]. So, for the given language, we can regard the adequacy of *both* systems to be strengthened by soundness and completeness theorems. In that case, the systems support each other. This would not be conclusive, of course, but it does help. In philosophy, we rarely get more.

I suggest that in some minds at least this mutual support is in fact enjoyed by current first-order logic. The semantic conception of logic is almost universally accepted, and the semantics of first-order logic is regarded by most as correct for such languages.[22] The rationalist conception of logic does not fare as well, however, It is a casualty of the prevailing anti-foundationalism. But, as evidenced by Wagner (1987), it is alive. For those who accept both, the soundness and completeness of first-order logic provide some assurance that we have it right on both counts. A good marriage.[23]

V. DIVORCE: LIFE WITHOUT COMPLETENESS

The above discussion concerns a single language, held fixed throughout and not questioned. We had a case in mind, of course, but the points are rather general. The harmony we found concerning first-order logic should not blind us to the possibility that this *language*, and its semantics, may not be adequate to codify actual mathematical practice (in its descriptive and deductive components). After all, both Aristotelian logic and propositional logic are complete,[24] but surely neither is adequate.

Some authors, myself included (1991), have argued that first-order languages and semantics are, in fact, not adequate. It can be conceded that first-order systems are sufficiently rich that the defects can be remedied by augmenting the object-language with non-logical terminology and axioms on a case-by-case basis (such as some set theory). But there is a limit to how far this can (or, at any rate, should) go. One alternative, higher-order logic, is inherently incomplete, in that there is *no* effective, sound, and complete deductive system for the semantics. Moreover, I suggest that *any* adequate language and semantics is incomplete.

There are, however, good, effective, deductive systems for second-order logic. One of them is a straightforward extension of a deductive system for first-order logic. Of course, it is not complete. The question at hand concerns the status of our two conceptions of logic under these circumstances. There are two possibilities: one can maintain both conceptions separately, or one can jettison the rationalist one.[25] I favor the latter, but a few remarks on each option are in order.

Joint custody

Suppose, then, that one has a model-theoretic semantics regarded as adequate for a language L, and one has a deductive system that is purported to codify ideal justification for that language. The deductive system is sound for the semantics, but not complete. I will not say more on how we might further support the adequacy of the semantics. But what of the deductive system and the rationalist conception of logic?

Under these circumstances, item (2b) above, stating that informal semantic validity entails ideal justifiability, would have to be regarded as very unlikely, if not false. As noted above, however, this is to be expected. It would take a fair amount of pre-established harmony for the human ability to justify inferences to match up to the semantic notion of validity. The converse (2a) can, and probably should, be maintained. If an argument can be justified, it is semantically valid.

Clearly, without completeness and (2b), we do not have the mutual support between the adequacy of the semantics and the adequacy of the deductive system. We can coherently maintain the *conformity* of the deductive system to rationalist standards (1a1). Once again, the axioms and rules of inference could be directly verified against our ability to detect self-evident justifications. But there seems to be no reason to accept the *sufficiency* of the deductive system (1a2). By hypothesis, one of the three premises used to establish it is false and another is unlikely at best. Under these circumstances, it is hard to imagine a form of argument that would establish sufficiency.

Of course, there could be inductive evidence for the sufficiency of a deductive system. It may be that it sanctions every inference we have considered so far that conforms to our intuitions of justification. Even in such a case, the best outlook would be to regard the sufficiency of the prevailing deductive system to be open. Unlike the scenario above, where completeness was assumed, one should admit and embrace the possibility that we may one day discover inferences that conform to rationalist standards but are not sanctioned by the deductive system. We might then augment the deductive system (but of course the new candidate would also be incomplete for the semantics). In short, the extension of the notion of "ideal justification" would be open-ended. This is rather consonant with the essential incompleteness of the logic.

To take an example, suppose it is known that an inference $\langle \Gamma, \Phi \rangle$ is semantically valid and that Φ cannot be proved from Γ in the deductive system. Both of these, of course, are established in the meta-theory (examples can be given for second-order logic). We might be tempted to quickly claim that Φ is, after all, justified on the basis of Γ (alone). Have we not *proved* that $\langle \Gamma, \Phi \rangle$ is valid? But this would conflate semantic validity and rationalist justification. The problem is that the *meta-theory* may have axioms or presuppositions which are not self-evident, or which otherwise do not conform to rationalist standards. Formal semantics is a theoretical, holistic, enterprise, not a rationalist one. Typically, it is formulated in set theory, and the proof that $\langle \Gamma, \Phi \rangle$ is valid may have substantial assumptions abut the "universe" of sets. So we cannot conclude from a proof that $\langle \Gamma, \Phi \rangle$ is semantically valid that Φ can be justified from Γ, at least not without further ado. However, we can, and should, hold open the possibility that we may one day discover a proof of Φ from Γ that does conform to rationalism.

Rationalism denied

Most of my views concerning higher-order logic are consistent with the outlook of the previous subsection. Indeed, I have very little to say on how one settles on a deductive system (in light of incompleteness), and it is conceded by all that pre-formal beliefs, or "intuitions," about inference are relevant to that enterprise. Nevertheless, I do not favor such an amiable divorce. The rationalist conception of logic is rejected altogether.

Let us consider the central notions of ideal justification and self-evidence. To pursue the above example, suppose it is known that $\langle \Gamma, \Phi \rangle$ is semantically valid, but that Φ cannot be proved from Γ in the currently favored deductive system. What would we do if we were given an *informal* proof of Φ from Γ that purports to conform to rationalist standards? Presumably, we would attempt to check whether it does so conform, whether it shows that Φ really is justified on the basis of Γ. What tools and measures are to be used in this check? Presumably, our intuitive judgments of correct inference and self-evidence. And there lies the problem.

I would suggest the possibility that our "intuitions" can simply run out at this point. Intuitive judgments may be based on paradigm examples, and may not pronounce on every case. If so, then we simply cannot judge on an *intuitive, pre-theoretical* level whether Φ really is justified on the basis of Γ. If an advocate of the rationalist conception accepted such a possibility, she might contend that the *lack* of judgment shows that Φ is after all *not* (yet) justified on the basis of Γ. In a sense, a lack of judgment is a negative judgment. On the other hand, the situation as described could equally be taken as indicating that the notion of self-evidence is *vague*. It has borderline cases, which might turn out to be borderline cases of justification itself. I leave it to rationalists to determine whether vagueness of the central notion of "correct inference" is consistent with the aims of the program.

There is, I think, a deeper problem, or at any rate a deeper difference between the rationalistic outlook and my own. According to the former, judgments of correct justification at the basic level are *prior* to theory, and thus *independent* of theory. Deductive systems are to codify these pre-existent and independent conceptions. Lacking a completeness theorem, semantics is pretty much irrelevant to this enterprise. I propose that, against this, justification is a holistic matter. Our post-theoretic judgments regarding correct inference are guided by our work in logical theory – both deductive and semantic. To borrow an overworked phrase, the "data," judgments of correct inference or "intuitions," are heavily laden with theory, at least at this point in history.

A rationalist might concede this, perhaps using a phrase like "corrupted by theory" instead of "laden with theory," but she would insist that the basic pre-theoretic notion survives. Otherwise, even the limited, retrenched foundationalist claims would be compromised. Even relative, ideal justification would be anchored in sand, not bedrock.

Consider, one more time, the situation in which it has been established (in the meta-theory) that a pair $\langle \Gamma, \Phi \rangle$ is semantically valid and that Φ cannot be proved from Γ in the deductive system. Suppose also that we have an informal proof of Φ from Γ. The question at hand is whether the informal proof shows that Φ is justified on the basis of Γ (alone). Call this situation T (for "theory"). Contrast it with another, call it NT (for "no theory"), which is identical except that there is no proof in the meta-theory that $\langle \Gamma, \Phi \rangle$ is semantically valid. That is, in NT, a (different) community knows that Φ cannot be proved from Γ in the deductive system, but has an informal proof, the same one as envisioned in situation T; and in this community, the question of whether $\langle \Gamma, \Phi \rangle$ is semantically valid is still open, or perhaps this community does not have a model-theoretic semantics at all.

We should assume that both communities are honest and straightforward in their efforts, and that both have the same *pre-theoretic* capacities for judging correct inference. The rationalistic assumption is that the community in situation T can, in effect, go back to their uncorrupted, pre-theoretical judgment, where ideal justification is a matter of self-evidence. It would undermine the rationalist enterprise if the semantical results of a powerful meta-theory were allowed to affect judgments of ideal justification, self-evidence, and the like. So, under these assumptions, the informal proof (in T and NT) is to be evaluated the same way in both situations, and both communities should come to the same conclusion *with the same confidence.* It is similar to two people measuring the length of one object using the same ruler under relevantly ideal conditions (assuming similar capacities to read rulers). Indeed, the only difference in the two situations here is the presence of the meta-theoretic proof (in T) that the inference is valid, and that is not relevant to the rationalistic conception of logic.

I do not claim that this scenario constitutes a *reductio ad absurdum* against the rationalist conception of logic, but at least for me the conclusion it points to is counterintuitive. It seems evident that the community in situation T would be more confident in the judgment that $\langle \Gamma, \Phi \rangle$ is correct, or that Φ can be justified on the basis of Γ (alone), simply because the informal proof is supported by a model-theoretic proof of semantic validity. That is, community T would be more willing to accept the informal proof, other things equal. In actual cases, we do rely on meta-theoretic results in judging correctness, or at least take them to be relevant (especially in set theory). In short, (semantic and deductive) theory guides our judgment. To be sure, we have an "intuitive" feel for correct argument, and we rely on this as well. Indeed, it is on intuitive judgment that meta-theory begins. However, I take it as uncontroversial that our intuitive judgments (of correct inference) are dynamic. They vary over time and are affected by successful theory, and education. The question at hand is whether there is a core of this intuitive judgment that survives all of the meta-theory, and remains prior to it. Even if there is such a core, can we now confidently "identify" it and separate it from the rest – the "corrupted" part of our intuitive judgment? Even more, is there any reason to attempt such a separation and build a philosophy of logic on it? I leave this part of the discussion with these questions, and my preference for negative answers.

Deductive systems without rationalism

A natural question at this point concerns the role of deductive systems, and criteria for evaluating them, on the semantic conception of logic. It might be appropriate for me simply to duck these questions by eschewing interest in deduction and deductive systems. It is not completely fair to avoid this question, however, especially in a volume on proof. Historically, the codification of deduction is a central task of logic, and it remains important in current studies. Proof theory has not been replaced by model theory. In making the semantic conception primary, it is necessary to at least establish a connection with traditional concerns.

As I see it, the purpose of developing a deductive system is cast in the same terms as that given by the rationalist. It is to codify the practice of giving proper (relative) justifications in mathematics. The difference, of course, is that I do not employ the same notion of "justification." What does that mean now?

Recall the last thesis in the list above:

(2b) If Φ is true under every interpretation of the language in which Γ is true, then Φ can be ideally justified on the basis of Γ.

In the context of that discussion, this is a substantive statement on the relationship between the semantic and the rationalist notions of correct inference, both regarded as autonomous. It was suggested that under these assumptions the thesis is most unlikely, especially in cases lacking a completeness theorem. Here, however, I have rejected the rationalist conception of logic and the autonomy of the notion of justification, however idealized.

A first attempt might be to use thesis (2b) (and its converse (2a)) as a *definition* of ideal justification. That would make this notion *completely* subordinate to semantics. In effect, it is the extreme opposite of rationalism. It simply eschews the notion of justification, replacing it with semantic validity. At least with regard to second-order logic, this too is counterintuitive. As pointed out by critics of second-order logic, there are cases in which the meta-theoretic *proof* that a given inference $\langle \Gamma, \Phi \rangle$ is semantically valid requires a substantial amount of set theory, enough that one can hardly say that Φ can be justified on the basis of Γ alone. An advocate of this subordinate notion of justification could retort that, in such cases, Φ is indeed justified on the basis of Γ alone, but one needs a substantial theory of sets in order to see this. After all, set theory is the vehicle for the semantics, the basis of justification so construed.

This is to bite a difficult bullet. There would be, for example, no standpoint from which one could *criticize* a deductive system, except by relating it to semantics. I prefer a middle course, in which we keep an "intuitive" notion of justification and use it to guide our theorizing, both semantic and deductive, and to evaluate deductive systems. There are several important differences between this notion of justification and its rationalist counterpart. There is no problem here with justification being laden with theory. The metaphor of the ship of Neurath (see, for example, Neurath 1932) applies to logic. The present notion of justification is admittedly vague. In at least some cases, there simply is no need to decide once and for all whether a given Φ really is justified on the basis of Γ. Borderline cases are to be expected when dealing with "intuitions" which are, after all, matters of training, experience, and temperament. Moreover, on such a view, there need not be a sharp boundary between logic and mathematics. Like any other science, the logic of mathematics may itself require a substantial amount of mathematics. The reason these differences are not problematic for me, and the reason that they do not entail extreme psychologism, is that no central foundational role is accorded to intuition. Logical theory, once again, is a holistic enterprise. It aims at precision, but does not have to start with something precise. Theory corrects as well as codifies our intuition.

VI. LOGIC AND COMPUTATION

It is widely believed that there is a close relationship between logic and computation. The historical roots of this run deep. For example, the Greek word for "syllogism" shares its etymology with "logistic," the theory of computation. In the early modern period, we have Hobbes' famous "By ratiocination, I mean computation" in *Concerning Body*, Part I, Ch. I (1839: 3), and from a different philosophical perspective, Leibniz' Universal Characteristic is a forerunner of mathematical logic (1686: XIV).

> What must be achieved is in fact this: That every paralogism be recognized as an *error* of *calculation*, and that every *sophism* when expressed in this new kind of notation ... be corrected easily by the laws of this philosophic grammar.... Once this is done, then when a controversy arises, disputation will no more be needed between two philosophers than between two computers. It will suffice, that, pen in hand, they sit down ... and say to each other: "let us calculate".

On the contemporary scene, it is no accident that recursive function theory, the study of computability as such, is a branch of *logic* and has its roots there. In Shapiro (1983), I argue that a principle motivation for the rigorous development of computability in the mid-1930s was to clarify the notion of an acceptable deductive system and, thus, to state Gödel's theorem in its full generality (and its ramifications for the Hilbert Program).

Along these lines, it is commonplace, and reasonable besides, to insist that deductive systems be effective in the sense that, in each case, the collection of correct (or sanctioned) inferences ought to be decidable, and thus the collection of sanctioned inferences (whose premises are finite) ought to be recursively enumerable. This, I believe, is reasonable, especially if the deductive system is to model or otherwise represent or codify the *process* of ideal justification, be the latter rationalist or otherwise. As suggested by Leibniz's metaphor, the system is to be *used* as a calculus to model *and check* actual informal reasoning (in the given language). Typically, deductive systems are constructed by specifying a collection of *immediate inferences* and defining the overall sanctioned inferences in terms of *deduction*, sequences of formulas (as above). If the deductive system is to serve its purposes, then a human should (ideally) be able to reliably determine whether a given inference is an immediate inference or not. It thus seems plausible to insist that there be an *algorithm* for determining whether a given inference is an immediate inference. Church's Thesis, the assertion that every algorithm computes a recursive function, is pretty much beyond dispute nowadays. It follows that the collection of immediate inferences of a deductive system should be recursive, and thus the collection of sanctioned inferences should be recursively enumerable.[26]

Conversely, for any recursively enumerable collection of inferences (whose premises are finite), there is a formal deductive system that "sanctions" all and only those inferences. Thus, a relationship between reasoning and computation amounts to a relationship between reasoning and formal deductive systems.

The sort of connection between computation and reasoning envisioned here has had its detractors. One of them is Brouwer, with his attacks on formalism and, for that matter, logic. Zermelo also rejected the connection, from almost the opposite perspective. This is not to mention opposition from general philosophy, notably the later Wittgenstein. These authors argue that in reasoning, one is not, or not merely, following a rule.

I believe that we are now in a position to shed some light on the relationship between computation and reasoning, and to delimit areas of genuine controversy. The issue, of course, is directly relevant to present concerns, especially the role of completeness.

Michael Detlefsen suggests a program for distinguishing two schools of thought concerning reasoning (see Chapter 8 in this volume). The "logic intensive" model focuses on deduction, and, in particular, on the *rules* one follows in reasoning correctly. The philosophical view behind this is that reasoning correctly is *constituted by* following the correct rules. The "intuition intensive" model rejects this idea. Reasoning is accomplished by some (other) faculty we possess; it is not simply the (blind) obedience to a rule. Detlefsen argues that Brouwer held a view like this (and herein lies his main objection to classical logical theory).

To mention the obvious, no one would go so far as to *identify* correct reasoning with computation (despite the above quote from Hobbes). Most algorithms do not correspond to, or represent, reasoning in any sense, even if we restrict attention to those that act on linguistic, or quasi-linguistic, items such as sentences or propositions. The thesis at hand is that correct reasoning is a *species* of computation. That is, if attention is restricted to a fixed language L, there is an algorithm that adequately represents correct reasoning in L. By the above equation, the thesis is that there is a particular formal system that adequately represents correct reasoning in L.

A central item in the present agenda is to articulate the notion of "representation" that is in place here. In what sense is it claimed that an algorithm or formal system "represents" reasoning? There are (at least) two interpretations. The strong thesis asserts that, in order to reason correctly, one must *grasp* and *execute* an appropriate algorithm, or equivalently, one must grasp and follow an appropriate formal deductive system. This is a form of the "logic intensive" view of reasoning. Indeed, on the strong thesis under consideration here, correct reasoning is literally constituted by following particular rules. The view also seems to underlie the rationalist conception of logic. As above, the rationalist holds that there is a fixed (*a priori*) notion of ideal relative justification. With this presupposition, a variation of the above argument that deductive systems must be effective suggests a thesis that correct reasoning is itself effective, and that there is a formal, effective deductive system that codifies ideal justification.[27]

There are deep problems, however, with this strong equation between reasoning and computation, quite independent of the above attack on rationalism. To sustain the thesis, we need an account of the notion of "grasping" that is in use when it is claimed that, to reason correctly, one must grasp a particular algorithm or deductive system. It cannot be a *conscious* grasping, in the sense that the grasper can *articulate* the relevant algorithm (as, for example, most of us can articulate an algorithm for addition). Most people who manage to reason correctly are not directly aware of any algorithm they may be following in the process. At the very best, only (some) logicians are consciously aware of the appropriate algorithm. There is also a problem concerning how natural language is rendered into the formal syntax of the algorithm or the formal language of the deductive system, and there is a problem of characterizing how the reasoner (correctly) applies the requisite algorithm. That is, are there to be rules for applying the "reasoning-algorithm"?[28]

The second, weaker articulation of the thesis connecting reasoning and computation is that, for a fixed language L, the collection of correct inferences (whose premises are finite) is recursively enumerable. That is, there is a formal system that *describes* the *extension* of correct reasoning in L. On this view, a reasoner does not have to be aware of and consciously execute such an algorithm, and, even more, one who does execute such an algorithm may not be reasoning. Moreover it may not be known, or even knowable, of any particular algorithm (or formal system) that it describes all and only the correct inferences in L. This version of the thesis is, I believe, consistent with the "intuition intensive" view of reasoning which, as it stands, does not say anything about the extension of "correct reasoning."

The weak thesis draws some support from mechanism and Church's Thesis. The relevant premises are that a human being is describable at some level as a complex machine, a mechanical system, and that the total output of any given mechanical system is recursively enumerable. But this is not enough to establish the thesis. The supposed algorithm is not to capture how *one* particular person *actually* reasons (even if we idealize on memory, attention span, etc.), but rather the extension of *correct* reasoning as such. Even if we allow that there is an effective way to separate out the portion of a person's "output" that is verbal and is reasoning, it would not follow that we can effectively delimit his *correct* reasoning. Thus, it seems that the argument from mechanism to our weak thesis needs another premise to the effect that it is possible for a human to reason correctly all the time. Even this is not sufficient. It is not enough that our idealized, mechanical person never make mistakes – dead people accomplish this much. The person must produce *all and only* correct inferences (in the language L). I suggest that even if we accept the mechanistic assumption and Church's Thesis, this further premise concerning a possible human super-reasoner does no more than beg the question. The idealized super-reasoner must still fall within the scope of mechanism. We have no reason to believe that there can be such a person unless we already believe that the extension of correct inference in L is recursively enumerable. Perhaps, as above, a rationalist has an argument for this, but what of the rest of us? Without a completeness theorem coupled to a good model-theoretic semantics for L, it is empty optimism to think that there is an effective formal system that not only "gets it right" but "gets all of it."

More importantly, the present (weak) thesis connecting reasoning and computation assumes that for the language L under consideration the extension of "correct inference in L" is fixed. That is, the statement that there is a formal system that describes the extension of correct inference presupposes that *there is* such an extension. To proceed any further, it is necessary to be more specific about the notion of "correct inference" that is in play here.

One possibility would be to revive the rationalist conception of logic and define "correct inference" in terms of the notion of ideal justification. Presumably, the extension of this is not only fixed, but fixed once and for all, *a priori*. As above, such a move would support the strong thesis, and the weak one as a corollary. But rationalism is not worth reviving.

A second possibility is to define "correct reasoning" in terms of semantic, model-theoretic validity. An inference is "correct" if and only if it is semantically valid. If this route is taken, then there is indeed a fixed extension for "correct inference" to the extent that there is a fixed semantics. That is, the semantics fixes the extension of "correct inference."

I would suggest, however, that under these circumstances we have no *a priori* reason to think that the extension is recursively enumerable. As Wagner (1987) points out, by opting for a *thoroughly* semantic conception of correct inference, we have divorced it from the reasoning abilities of humans, however idealized. It would be fortuitous if validity ended up being recursively enumerable (as it does with first-order languages), but we cannot *start* with this assumption. We start with semantics.

The best course, I believe, is the intermediate one sketched in the last subsection. The semantic conception of logic is adopted, but the notion of correct inference is not identified with semantic validity. On this view, ideal justification is *not* regarded as fixed, at least not fixed once and for all. There are two senses to this. First, the notion of ideal justification, as I see it, is *dynamic*.[29] The extension of "correct inference" is molded by successful theory, both semantic and deductive. To accommodate this, one can fix attention on a fixed time t, and consider the notion of ideal-justification-at-t, those inferences that are justified by resources available at time t. Our weak thesis is now the assertion that there is an algorithm or formal system that describes correct-inference-in-L-at-t. Second, the notion of ideal justification (at time t) is, or may be, vague – there very well might be borderline cases of correct reasoning. If so, our weak thesis loses much of its force, if not its sense. Surely, an algorithm or formal system cannot *exactly* describe the extension of a vague concept (if there is such a thing). No doubt, one can still hold that the inferences sanctioned by a given formal system more or less coincide with correct-reasoning-in-L-at-t. Of course, this is another retrenchment. It is, I believe, eminently plausible. Indeed, this thesis is a presupposition, or regulative ideal, of proof theory. We would not even try to develop formal deductive systems if we did not believe that a reasonable model could be achieved. But our retrenched weak thesis is a far cry from the original equation between reasoning and computation.

NOTES

*This essay was written, in part, while I held a Faculty Professional Leave from The Ohio State University, a Fellowship for College Teachers and Independent Scholars from the National Endowment for the Humanities, and a visiting Fellowship at the Center for Philosophy of Science at the University of Pittsburgh. I am grateful to all three institutions. The project has benefited from comments and discussion from many people. Included are George Boolos, Michael Detlefsen, Nicolas Goodman, Jaakko Hintikka, Robert Kraut, Lila Luce, Penelope Maddy, Charles McCarty, Gregory Moore, Michael Resnik, Barbara Scholz, George Schumm, Allan Silverman, and Robert Turnbull.

1 It might be added that connections between fields go both ways. If misgivings about real analysis persist, one can doubt the coherence of our talk of sets or sequences. The foundational work shows that, in a sense, real numbers are equivalent to sets of natural numbers.
2 A list of twentieth-century foundationalist programs must, of course, include intuitionism. Whatever the fate of other efforts, it is clear that intuitionism will be alive in the next century. It will be pursued, not just talked about. Here, however, it would take us too far afield, partly because intuitionism goes in a direction radically different from the others and partly because it is revisionist. Present focus is on the foundations of classical mathematics, programs that leave the beast (more or less) as they find it.
3 In addition to van Heijenoort (1967) and Goldfarb (1979), the distinction between a universal language and a formal language, between "logic-as-language" and "logic-as-calculus," is discussed in more detail in Hintikka (1988) and Cocchiarella (1988). Resnik (1980) traces the discussion at cross-purposes in the correspondence between Frege and Hilbert to their different orientations.
4 There are no nontrivial first-order languages that lack nonlogical terminology. In first-order cases, all predicates and relations (except identity) are nonlogical. Higher-order systems, on the other hand, have predicate and relation variables. Thus, all of the logicist systems are higher-order.
5 There are exceptions. See Detlefsen (1986).
6 The role of effectiveness in deductive systems is taken up below.
7 Kitcher (1983) argues, for example, that the Cauchy–Weierstrass foundation of analysis, the familiar $\varepsilon-\delta$ formulations of limit, continuity, etc., were accepted only because the previous techniques involving infinitesimals, divergent sequences and the like were breaking down in practice. People were no longer sure whether a given process, or inference, would lead to correct results. At least this time, the rigor

of the Cauchy-Weierstrass definitions was needed on internal grounds. grounds.

8   See Goldfarb (1988) for a lucid presentation of Poincaré's contribution to the development of logic and foundations of mathematics.
9   At least part of the Fregean goal was to rid (some) mathematics of Kantian intuition. The proposed move would not necessarily reintroduce it, since the postulated faculty need not be directly related to the structure of sensory perception. The nature of intuition is a recurring theme in Parsons (1983).
10 Compare this to the so-called "gap" in Euclid's elements that, if a line goes through a circle, it must intersect it somewhere.
11 As noted above, logicism denies value to *any* formal meta-theory. The theory to be used without the benefit of more theory is the very first one.
12 There are exceptions. Many treatments of first-order logic contain "logical axioms," and the ω-rule only applies to arithmetic.
13 In this respect, sufficiency and conformity are the same kind of thing as Church's Thesis.
14 This assumes that the pre-formal notion of real justification is "transitive." If Φ is justified on the basis of $\Gamma 1$ and every member of $\Gamma 1$ is justified on the basis of $\Gamma 2$, then Φ is justified on the basis of $\Gamma 2$. In particular, no finite bound is placed on the "length" of a justification.
15 This phrase is often associated with extreme psychologism, in which case it is not a fair characterization of some of Frege's predecessors, notably the algebraists Boole, Peirce, and Schröder. This is a story that cannot be pursued here.
16 Hintikka (1988) argued that Hilbert held a semantic, or "model-theoretic" conception of logic. This, I believe, is clear for Hilbert's relatively early period, when he corresponded with Frege. The later Hilbert Program, however, was more concerned with deduction than with interpretation, with proof theory rather than model theory. Even so, the program prescribed formal languages, languages that might as well be uninterpreted. This much, at least, is compatible with the notion of different interpretations of the language, and thus with the semantic conception of logic. Thus, I agree that the Hilbert Program need not be seen as a rejection of the semantic conception of logic. Proof theory can augment model theory; it need not supplant it.
17 The possibility that terminology for natural numbers is not to be interpreted differently in different models, that it be "logical terminology," has been studied. It is called ω-logic. It follows from the compactness theorem that in first-order systems the "assumption" that 0, $s0$, $ss0$, etc. are all of the natural numbers cannot be stated. In a second-order system, of course, it can be. Thus, in a sense, ω-logic lies between first-order logic and second-order logic (see Shapiro 1991). It is fruitful to think of the difference between first-order logic and higher-order logic as a difference concerning the border between logical and nonlogical terminology.
18 This question of semantic sufficiency is similar to one in Kreisel (1967) concerning the relationships between what he calls "informal

validity" and what may be called "set-theoretic validity." His remarks are limited to the context of common nth order languages, and he focuses on single formulas: a formula Φ is *informally valid* if it is true under *all* interpretations of the language, and Φ is *set-theoretically valid* if Φ is true under all interpretations of the language *whose domain is a set*. The principle that informal validity is co-extensive with set-theoretic validity is treated (and expanded) in Shapiro (1987).

19 Questions like this are central in Kreisel's (1967) analysis of informal rigor.
20 I take "Γ is true" as an abbreviation of "every member of Γ is true."
21 To get frivolous about this exercise, it might be noted that soundness follows from (1a1), (2a), and (1b1); and completeness follows from (1b2), (2b), and (1a2). But soundness and completeness are formal, mathematical matters, internal to the developed structures.
22 Exceptions include intuitionists, nominalists (who reject talk of abstract models and interpretations altogether), and strict finitists (who reject the infinite, without which the completeness theorem cannot be proved).
23 There is an interesting analogy with Church's Thesis, which is supported by the fact that several different (independent) attempts to codify computability all yield the same extension. See Shapiro (1981).
24 For Aristotelian logic, see Corcoran (1972); for propositional logic, see any competent textbook in symbolic logic.
25 One could also jettison the semantic conception and maintain the rationalist one. This would be a return (of sorts) to logicism. Since my arguments in favor of second-order logic are "semantic" (broadly conceived), it would be foolhardy of me to reject the semantic conception. But one person's proof is another's *reductio ad absurdum*.
26 A stronger version of Church's Thesis is that if a human can be trained to reliably determine membership in a given set, then the set is recursive. This, in effect, combines two steps in the above argument.
27 See Wagner (1987) for such an argument. Wagner calls a logic "complete" if its consequence relation is effectively enumerable. It is more common, however, to use the term "complete" to designate a relation between a formal deductive system and a model-theoretic semantics, as above.
28 This, of course, is a Wittgensteinian problem. See Wright (1980) and Blackburn (1984).
29 For a more extensive discussion of the dynamic nature of "correct inference," see Shapiro (1989).

REFERENCES

Blackburn, S. (1984) *Spreading the Word*, Oxford: Clarendon Press.
Bolzano, B. (1817) *Rein analytischer Beweis des Lehrsatzes, dass zwischen je zwei Werthen, die ein entgegengesetztes Resultat gewaehren, wenigstens eine reelle Wurzel der Gleichung leige*, Prague: Gottlieb Haase.

Boolos, G. (1975) "On second-order logic," *Journal of Philosophy* 72: 509–27.
Cocchiarella, N. (1988) "Predication versus membership in the distinction between logic as language and logic as calculus," *Synthese* 77: 37–72.
Coffa, A. (1982) "Kant, Bolzano, and the emergence of logicism," *Journal of Philosophy* 79: 679–89.
Corcoran, J. (1972) "Completeness of an ancient logic," *Journal of Symbolic Logic* 37: 696–702.
Detlefsen, M. (1986) *Hilbert's Program*, Dordrecht: D. Reidel.
Goldfarb, W. (1979) "Logic in the twenties: the nature of the quantifier," *Journal of Symbolic Logic* 44: 351–68.
—— (1988) "Poincaré against the logicists," in W. Aspray and P. Kitcher (eds) *History and Philosophy of Modern Mathematics*, Minnesota Studies in the Philosophy of Science, vol. 11, Minneapolis, MN: University of Minnesota Press, 61–81.
van Heijenoort, J. (1967) "Logic as calculus and logic as language," *Synthese* 17: 324–30.
Hilbert, D. (1925) "Über das Unendliche," *Mathematische Annalen* 95: 161–90.
Hintikka, J. (1988) "On the development of the model-theoretic viewpoint in logical theory," *Synthese* 77: 1–36.
Hobbes, T. (1839) *Concerning Body*, in W. Molesworth (ed.) *The English Works of Thomas Hobbes*, vol. 1, London: John Bohn.
Kitcher, P. (1983) *The Nature of Mathematical Knowledge*, New York: Oxford University Press.
Kreisel, G. (1967) "Informal rigour and completeness proofs," in I. Lakatos (ed.) *Problems in the Philosophy of Mathematics*, Amsterdam: North-Holland, 138–86.
Leibniz, G. (1686) "Universal science: characteristic XIV, XV," *Monadology and other Philosophical Essays*, trans. P. Schrecker, Indianapolis, IN: Bobbs-Merill, 1965 11–21.
Neurath, O. (1932) "Protokollsätze," *Erkenntnis* 3: 204–14.
Parsons, C. (1983) *Mathematics in Philosophy*, Ithaca, NY: Cornell University Press.
Poincaré, H. (1909) "La Logique de l'infini," *Revue de Métaphysique et Morale* 17: 461–82.
Resnik, M. (1980) *Frege and the Philosophy of Mathematics*, Ithaca, NY: Cornell University Press.
—— (1985) "Logic, normative or descriptive? The ethics of belief or a branch of psychology," *Philosophy of Science* 52: 221–38.
Shapiro, S. (1981) "Understanding Church's thesis," *Journal of Philosophical Logic* 10: 353–65.
—— (1983) "Remarks on the development of computability," *History and Philosophy of Logic* 4: 203–20.
—— (1985) "Second-order languages and mathematical practice," *Journal of Symbolic Logic* 50: 714–42.
—— (1987) "Principles of reflection and second-order logic," *Journal of Philosophical Logic* 16: 309–33.

—— (1991) *Foundations Without Foundationalism*, Oxford: Oxford University Press.

—— (1991) *Foundations Without Formalism*, Oxford: Oxford University Press.

Skolem, T. (1952) "Some remarks on the foundation of set theory," *Proceedings of the International Congress of Mathematicians, Cambridge, Massachusetts, 1950*, Providence, RI: American Mathematical Society, 1952, 695–704.

Tait, W. (1981) "Finitism," *Journal of Philosophy* 78: 524–46.

Tarski, A. (1935) "On the concept of logical consequence." Reprinted in A. Tarski *Logic, Semantics and Metamathematics*, Oxford: Clarendon Press, 1956, 417–29.

Wagner, S. (1987) "The rationalist conception of logic," *Notre Dame Journal of Formal Logic* 28: 3–35.

Weyl, H. (1987) *The Continuum: A Critical Examination of the Foundations of Analysis* (trans. by S. Pollard and T. Bole), Kirksville, MO: Thomas Jefferson University Press.

Wright, C. (1980) *Wittgenstein on the Foundations of Mathematics*, Cambridge, MA: Harvard University Press.

8

BROUWERIAN INTUITIONISM*

Michael Detlefsen

SUMMARY

The focal question of this essay is what if any role logical inference should be taken to play in mathematical reasoning. Special attention is given to the idea of Brouwer and Poincaré that, by its very topic-neutral character, logical inference is unsuitable for use in mathematical reasoning, and the epistemology standing behind this idea is partially developed. According to this epistemology, intuition is needed not only to supply axioms, but also to power inference.

I. PRÉCIS

The aims of this essay are twofold: first, to say something about that philosophy of mathematics known as "intuitionism" and, second, to fit these remarks into a more general message for the philosophy of mathematics as a whole. What we have to say on the first score can, without too much inaccuracy, be compressed into two theses, the first being that the intuitionistic critique of classical mathematics can be seen as based primarily on epistemological rather than on meaning-theoretic considerations, and the second being that the intuitionist's chief objection to the classical mathematician's use of logic does *not* center on *the use of particular logical principles* (in particular, the law of excluded middle and its ilk), but rather on the *role* the classical mathematician assigns (or at least extends) generally (i.e. regardless of the *particular* principles used) to the use of logic in the production of mathematical proofs. Thus, the intuitionist critique of logic that we shall be presenting is far more radical than that which has commonly been presented as the "intuitionist critique."

On the second, more general, theme, what we have to say is this: some restriction of the role of logical inference in mathematical proof such as that mentioned above is necessary if one is to account for the seeming difference in the epistemic conditions of provers whose reasoning is based on genuine insight into the subject-matter being investigated, and would-be provers whose reasoning is based not on such insight, but rather on principles of inference which hold of every subject-matter indifferently. Poincaré urged this point repeatedly, but, in the rapid development of logic in this century, it seems to have been forgotten. I think it deserves more attention than it has received and that, when properly taken into account, it provides an interesting "new" ground for a mathematical epistemology sharing many of the features of Brouwerian intuitionism.

Poincaré's insight suggests an epistemology which operates according to a principle of epistemic conservation: there can be no increase in genuine knowledge of a specific mathematical subject without an underlying increase in subject-specific insight into (i.e. *intuitional* grasp of) that subject. Thus, the need for intuition cannot be avoided in mathematics even if it has supplied a set of axioms. Hence, purely logical inference cannot add to our genuinely mathematical knowledge, and thus cannot be given a very important role in proof.

This, in brief, is the position to be developed in this essay. As sensitive to Poincaré's concern regarding the plausibility of a mathematical epistemology which allows increases in mathematical knowledge without a correspondingly increased insight into the particular mathematical subject involved. This new (i.e. non-classical) epistemology requires a new conception of inference, for in order for a truth to be proven it requires that it be "experienced" in a certain way. And this new conception of inference severely restricts the role of logical inference in proof. By (classical) logical analysis or inference, one can extract all kinds of propositions from a given experienced proposition. But only some of these extracted propositions are themselves "experienceable" in the appropriate way (just as, in the case of empirically perceived truths, only some of their logical consequences are themselves empirically perceivable). And none of them are experienced in the appropriate way solely by their being shown to be related to the premises by logical means.

This, in brief, is the position to be developed in this essay. As mentioned, it seeks to present mathematical intuitionism as essentially an epistemological rather than a meaning-theoretic view. It also seeks to distance it from the solipsism commonly attributed to Brouwer, and to focus instead on Poincaré's concern over the place of purely logical inference in genuinely mathematical reasoning. The result, we hope, is both an interesting way of thinking about intuitionism, and a renewed appreciation of the importance of Poincaré's point for the philosophy of mathematics.

II. POINCARÉ'S CONCERN

Poincaré presented his point in the form of an observation which he then put forth as a central "datum" for the philosophy of mathematics. The substance of this observation is quite simple, and can be presented as the result of a thought-experiment to the following effect.

> Imagine two cognitive agents M and L. M has the kind of knowledge or understanding of a given mathematical subject S that we typically associate with the master mathematician. L, on the other hand, has the sort of epistemic mastery of S that is typical of one whose epistemic command of S consists in a knowledge of a set of axioms for S plus an ability (possibly superb) to manipulate or process those axioms according to acknowledged *logical* means. Query: Is there any significant difference between the epistemic condition of M and the epistemic condition of L *vis-à-vis* their status as mathematical knowers?

In Poincaré's view, the answer is "Yes." Even perfect *logical* mastery of a body of axioms would not, in his view, represent genuine mathematical mastery of the mathematics thus axiomatized. Indeed it would not in itself be indicative of any appreciable degree of mathematical knowledge at all: knowledge of a body of mathematical propositions, plus mastery over their logical manipulation, does not amount to mathematical knowledge either of those propositions or of the propositions logically derived from them.

On Poincaré's view, then, genuine mathematical reasoning does not proceed in "logic-sized" steps, but rather in bigger steps – steps requiring genuine insight into the given mathematical subject being inferentially developed. This sets it at odds with logical reasoning which, *by its very topic-neutral character*, neither requires nor even admits use of such insight in the making of inferences. In thus foreswearing all appeal to information that derives from the particularities of the specific subject-matter under investigation, logical reasoning also foreswears the easy, loping stride of one familiar with the twists and turns of a given local terrain, and opts instead for the halting step of one who is blind to the special features of all localities, and who must therefore take only such steps as would be safe in *any*. In Poincaré's view, the security thereby attained cannot make up for the blindness which it reflects. Logical astuteness may keep one from falling into a pit, but having a cane with which to feel one's way is a poor substitute for being able to see.

It was from this point of view that Poincaré framed his criticism of the "logicians" (e.g. Couturat, Frege, Peano, and Russell), a criticism which occupied a place of fundamental importance in his overall philosophy of mathematics.

> The logician cuts up, so to speak, each demonstration into a very great number of elementary operations; when we have examined these operations one after the other and ascertained that each is correct, are we to think we have grasped the real meaning of the demonstration? Shall we have understood it even when, by an effort of memory, we have become able to reproduce all these elementary operations in just the order in which the inventor had arranged them? Evidently not; we shall not yet possess the entire reality; that I know not what, which makes the unity of the demonstration, will completely elude us ...
>
> If you are present at a game of chess, it will not suffice, for the understanding of the game, to know the rules of moving the chess pieces. That will only enable you to recognize that each move has been made comformably to these rules, and this knowledge will truly have very little value. Yet this is what a reader of a book on mathematics would do if he were a logician only. To understand the game is wholly another matter; it is to know why the player moves this piece rather than that other which he could have moved without breaking the rules of the game. It is to perceive the inward reason which makes of this series of moves a sort of organized whole. This faculty is still more necessary for the player himself, that is, for the inventor.
>
> Poincaré (1905: 217–18)

The very possibility of the science of mathematics seems an insoluble contradiction. If this science is deductive only in appearance, whence does it derive that perfect rigor no one dreams of doubting? If, on the contrary, all the propositions it enunciates can be deduced one from another by the rules of formal logic, why is not mathematics reduced to an immense tautology? The syllogism can teach us nothing essentially new, and, if everything is to spring from the principle of identity, everything should be capable of being reduced to it. Shall we then admit that the enunciations of all those theorems which fill so many volumes are nothing but devious ways of saying A is A?

Without doubt we can go back to the axioms, which are the source of all these reasonings. If we decide that these cannot be reduced to the principle of contradiction, if still less we see in them experimental facts which cannot partake of mathematical necessity, we have yet the resource of classing them among synthetic *a priori* judgements. This is not to solve the difficulty, but to baptize it; and even if the nature of synthetic judgements were for us no mystery, the contradiction would not have disappeared, it would only have moved back; syllogistic reasoning remains incapable of adding anything to the data given in it; these data reduce themselves to a few axioms, and we should find nothing else in the conclusions.

No theorem could be new if no new axioms intervened in its demonstration; reasoning could give us only the immediately evident verities borrowed from direct intuition; it would be only an intermediary parasite, and therefore should we not have good reason to ask whether the whole syllogistic apparatus did not serve to disguise our borrowing? ...

If we refuse to admit these consequences, it must be conceded that mathematical reasoning has of itself a sort of creative virtue and consequently differs from the syllogism.

The difference must even be profound. We shall not, for example, find the key to the mystery in the frequent use of that rule according to which one and the same uniform operation applied to two equal numbers will give identical results.

All these modes of reasoning, whether or not they be reducible to the syllogism properly so called, retain the analytic character, and just because of that are powerless.

Poincaré (1902: 31)

To bolster his general distinction between logical and mathematical reasoning, Poincaré offered an example; a case of reasoning which he took to be paradigmatic of genuinely mathematical reasoning, and which at the same time he believed to be nonlogical (or, to use his term, "non-analytical") in character, namely, mathematical induction.

After giving several illustrations of the importance of mathematical induction to mathematics, Poincaré turns to the issue of its character, arguing that it is synthetic, rather than analytic, since its conclusion "goes beyond" its premises rather than being a mere restatement of them "in other words." At the same time, however, he argues that it is entirely rigorous, and so, rightly classified as *mathematical* reasoning. It is precisely this combination of characteristics – syntheticity and rigor – that Poincaré takes to typify genuine mathematical reasoning. And it is the first of these – syntheticity – which he calls upon to distinguish mathematical from logical reasoning.

This, in outline, is Poincaré's view of what are the epistemologically important differences between logical and mathematical reasoning. We do not, however, propose to discuss it in detail here. In particular, we intend no defense either of its understanding of the analytic/synthetic distinction, or of its contention that that distinction, thus understood, provides the correct means of explaining the more basic conviction that there is an important difference between the epistemic conditions of the purely logical reasoner and the mathematical reasoner *vis-à-vis* their mathematical knowledge. Rather, it is this more basic conviction which is of chief interest to us, since we take it to constitute the general problematic that is basic to Brouwer's intuitionism and, in particular, his critique of classical mathematics.

Taken seriously, this problematic promises to have some important effects on one's conception of mathematical knowledge. One such effort is that of implying what might be called a *modal* – as opposed to a *subjectival* – construal of mathematical knowledge. On the subjectival construal, the typology of knowledge follows a classification scheme which sorts knowledge according to the subject-matter of its content. Thus, in order for one's knowledge that **p** to count as κ-knowledge (i.e. knowledge of type κ), all that is required is that **p** be a truth of *subject-matter* κ. On such a model, mathematical knowledge becomes simply knowledge of a *mathematical* truth (i.e. knowledge of a truth belonging to a mathematical subject-matter).

On what we are calling the *modal* conception, on the other hand, the typology of knowledge does not follow a subject-matter classification of the propositions known. It marks as well certain differences in the particular cognitive attitude taken. Thus, mathematical knowledge becomes more than simply knowledge of a mathematical proposition, and is distinguished by a certain *mode* or *kind* of cognitive state as well.

It is, of course, a difficult question to say what exactly it is that is to distinguish that special mode of knowledge known as mathematical knowledge from other modes of knowledge. Indeed, there is room for dispute over this within the modalist camp. Poincaré, for example, believed that it had primarily to do with one's ability to see the role or position that a given proposition plays in the larger subject to which it belongs; so that one comes to know something mathematically by having a global vision of the place of that proposition within some larger epistemic enterprise. Brouwer, on the other hand, believed that to mathematically know a truth was to "experience" it in a certain way. Others would say that the distinctive feature of mathematical knowing is its freedom from empirical considerations. Still others would insist that degree-of-certainty plays an important role. And so on.[2]

We shall make no attempt to decide between such rival modalist epistemologies here, since the implications of the modal conception of mathematical knowledge with which we are principally concerned are of a more general character than those pertaining to some particular articulation of it. Of particular importance to us in this connection are certain implications regarding how we are to conceive of the *growth* or *extension* of mathematical knowledge under a generally modalist mathematical epistemology. And, as we shall see shortly, the use of logical inference in the production of mathematical knowledge is only compatible with such weak modalist conceptions as take relatively large-scale, coarsely differentiating features (e.g. high degree of certainty or *a priority*) as the distinguishing features of mathematical knowledge.

As already noted, the key idea of the modalist conception is that to have mathematical knowledge of a given proposition **p** is to have a certain *kind* of knowledge that **p**. Thus, if a given kind of knowledge of **p** is to be extended to another proposition **q** by means of an inference from **p** to **q**, then that inference must preserve the special characteristics of knowledge of **p** that are responsible for its being of that kind. Therefore, if knowledge of a particular kind κ is to be extended by means of *logical* inference, then logical inference must preserve those features of a given piece of knowledge that make it κ-knowledge. To put it another way: if κ-knowledge of **p** is to be extended to κ-knowledge of **q** by means of a logical inference from **p** to **q**, then the distinguishing features of κ-knowledge must be included among those properties of beliefs that are preserved by logical inference.

This constraint is not trivial or powerless, since there clearly are types of knowledge that are distinguishable by features that are not preserved by logical inference. As a specific example of this, let us consider knowledge by direct sensory experience. I look at the grass outside my window and see that it is green. Turning my chair in the opposite direction, I view the carpeting in the hallway and see that it is grey. *Logically,* I can infer from the knowledge thus obtained that the grass outside my window is green and the carpeting in the hallway is grey. However, owing to the practical difficulties involved (e.g. my inability to direct my eyes in opposite directions at one and the same time, to see around or through corners, etc.), I cannot produce a direct sensory experience whose content is that the grass outside my window is green and the carpeting in the hallway is grey. Thus, logically extending the content of knowledge gained by direct sensory experience does not guarantee that the content thus extended will be accessible via *the same cognitive mode* (in this case, direct sensory experience).

A different, though equally mundane, kind of example can be found by considering such processes as ordinary counting. A ticket-taker at a basketball game knows, by having (partially) counted them as they entered the gate, that there are at least 25 people seated in his section. In order to determine how the people seated in his section are distributed over lower versus upper arena seats, he decides to count those seated in upper arena seats. He counts zero people seated there. He thus knows by ordinary counting that there are at least 25 people seated in his section and that there is no one seated in an upper arena seat. From this it logically follows that there are at least 25 people seated in the lower arena seats. It is not true, however, that this is known by an ordinary (partial) counting of those occupying lower arena seats, since the ticket-taker arrived at his conclusion without having actually counted (in what we are calling the "ordinary sense") the occupants of the lower arena seats.

What has just been said of knowledge by direct perception and knowledge by ordinary counting can also be said of other types of knowledge. Indeed, as we shall see later, it applies specifically to Brouwer's conception of mathematical knowledge, which he takes to be constituted by a kind of "experience" (or intuition). In each case, the crucial issue is whether a given kind of cognitive mode (direct sensory perception, ordinary counting, Brouwerian mathematical intuition, etc.) can be manipulated or controlled in such a way as to be guaranteedly reproduced at all the propositions that are logically derivable from a proposition describing the content of a given such state. And, in each case, the answer is "no." We do not have the practical capacity to manipulate the having of such kinds of mental states in the full range of ways that we can *logically* manipulate their *contents*. Therefore, logical inference does not preserve cognitive mode.

Logical manipulation of the content of a mental state is thus one thing, and *practical manipulation* of its cognitive mode another. Therefore, the assumption that mathematical knowledge is extendable by logical reasoning is not an innocent one. The only clear capacity of logical inference is that of an abstractive device; that is, a device for *separating off* the *content* of a given cognitive state from its *mode of occurrence*, and submitting that content to various sorts of analyses which issue in the production of new contents. As such, it is not automatically an extension or continuation of the cognitive state from which the content was separated, but rather a focused *reflection on* its content. Being thus focused on content rather than on cognitive mode, it may be expected to carry forward the content of a given piece of knowledge, but without any corresponding guarantee that the content thus forwarded occurs in the same cognitive mode as the original.

Such failure to extend the cognitive mode of a given piece of knowledge is, of course, no tragedy if occurrence in that mode is incidental, or at least inessential, to its overall epistemic character and/or value. And this may be the case in some of our examples (e.g. knowledge by ordinary counting). Brouwer and his "pre-intuitionist"[3] predecessor Poincaré did not, however, believe that it is so in the case of mathematical knowledge generally, and their belief was rooted in that observation which we are referring to here as "Poincaré's Concern": namely, that the epistemic condition of one who has gained a logical or axiomatic mastery over a given mathematical subject is inferior to that of one who has a genuine mathematical mastery of it.[4] They had different ways of accounting for this difference, but both clung to it as a basic fact of mathematical epistemology. In this paper, we shall consider only Brouwer's epistemology, leaving Poincaré's for another occasion. However, as a preparation for presenting Brouwer's ideas, it will prove useful to at least lay out the rudiments of its chief antithesis; namely, the so-called "classical epistemology."

III. CLASSICAL EPISTEMOLOGY

The epistemology underlying classical mathematics (which, for brevity, we shall refer to as *classical epistemology*) emphasizes the *contentual* ingredient of knowledge, and de-emphasizes the matter of its *cognitive mode*. According to it, the mathematical knower may rely on introspective experience (or intuition) of some sort to arrive at the initial propositions of his epistemic edifice, but from that point on he is free to abstract away from or ignore the *noncontentual* aspects of that experience, and concentrate instead on its *contentual* component. What, on this account, is of primary epistemic importance concerning the cognitive mode of a given epistemic event is what might be called its *credential effect*; that is, the degree of certitude it confers upon the proposition expressing its content. But since widely different cognitive modes are capable of having the same credential effect, identifying the epistemic significance of a cognitive mode with its credential effect produces an epistemology which tends to reduce the number of epistemically significant differences between cognitive modes. This, in turn, leads to a view of inference which sees it as having relatively little obligation to preserve the features of the cognitive mode of the premises (since so few of them are of any epistemic significance). Beyond credential effect, classical mathematical epistemology, at least in some of its variants, may make room for such large-scale characteristics of cognitive mode as its *aprioricity/aposterioricity*. However, sensitivity to such large-scale features will surely not provide a grid fine enough to make the sorts of small-scale demarcations – in particular, the sort of demarcation between logical and mathematical reasoning described in the preceding section – that Brouwer and Poincaré regarded as being of prime importance to a well-developed mathematical epistemology.

On the classical view, then, proof or inference is a procedure of the following sort: the mathematical knower begins with knowledge occurring in a certain cognitive mode; he then abstracts away from all characteristics of that cognitive mode that he regards as epistemically irrelevant, leaving him with only such of its features as credential effect and, say, *aprioricity/aposterioricity* to attend to; having thus narrowly restricted the focus of his epistemic concern, he has correspondingly widened the horizons of inference by making it possible to extend his knowledge to any *new* proposition which can be obtained from the proposition expressing the content of his old knowledge by means capable of preserving (sufficiently much of) his narrowly restricted focal epistemic desiderata (namely its credential effect and *aprioricity/aposterioricity*).

This relationship between the narrowing of the range of the epistemically significant features of cognitive mode and the corresponding broadening of the inferential horizon deserves a further word of elaboration. For it is really not the narrowing of the range of epistemically significant features of cognitive mode *per se*, but rather the *particular* narrowing to the likes of credential effect and *aprioricity* that produces the corresponding widening of the range of possible inferences: such epistemic attributes as credential effect and *aprioricity* are preserved by a wider range of inferential transformations than are such more fine-grained attributes as direct sensory perceived-ness or Brouwerian mathematical intuited-ness. In identifying the crucial feature(s) of warrantedness with properties that are so little dependent on the more fine-grained characteristics of cognitive mode, it increases its "liquidity" or transferability by decreasing the extent to which the cognitive mode of a properly inferred conclusion must resemble that of the premise(s) from which it is inferred.

Thus it is that the less (more) stringent the demands on preservation of the features of the cognitive mode of a premise are, the less (more) restricted are the opportunities for inference. In placing relatively weak demands on the preservation of cognitive mode, classical epistemology thus leaves a correspondingly greater role for inferential justification. This is perhaps its most significant point of contrast with Brouwerian epistemology. As we shall see in the next section, the demands on preservation of cognitive mode coming from Brouwerian epistemology are so strong as to leave very little opportunity for turning a justification for one proposition into a justification for another. And since the ability to use a justification for one proposition to produce a justification for another seems to be the essence of inferential justification, the result is that Brouwerian epistemology leaves comparatively little scope for inferential justification – thus the comparatively greater need for what might be called "intuition."

On the above analysis, then, logical inference (by which mathematical knowledge is to be extended) is essentially a comparative reflection on *contents*, where these contents are taken to be relatively independent of the epistemic processes or activities to which they are attached. It does not *reflect* or *express* the characteristics of the epistemic activities underlying those contents in such a way as to force logical relations to imply practical relations between them. The basic idea of classical epistemology is thus that the epistemically relevant characteristics of a given experience or piece of intellectual activity are separable or detachable from it. Memory, or some like capacity, is called upon to sustain the epistemic effects and potency of a given piece of mental activity long after the activity itself has ceased to exist in experience. That memory-like capacity functions to "retain" the content and warrant of an experience (or other warranting activity) so that it can be passed on to propositional contents not occurring in that cognitive mode. *Logical* analysis then "detaches" the contentual results of epistemic processes from those processes themselves, and treats them as independent entities, the result being that logical inference or knowledge is taken to consist in a manipulation of *warrant contents* rather than of *warranting processes*.

In classical epistemology, then, the epistemic effect of a warrant is quite stable – being preserved under transformations that allow the characteristics of the particular process(es) that originally produced the warrant to be greatly altered.

Motivating this classical epistemology of inference is a certain conception of language and of the epistemological enterprise generally that we shall call the *logic-intensive* or *representation-intensive* view. The basic idea behind this view is that, though knowledge may perhaps begin with "intuition" or experience of some kind, it nonetheless can, and in many cases must and should, be extended without a corresponding extension of that intuition or experience. Thus, though experience may be necessary in order for knowledge to *begin*, it has strictly limited value as a means of *extending* knowledge.[5]

It is this belief in the limited epistemic exploitability of experience that forms the basis of the logic-intensive or representative-intensive view. It takes experience itself to be a relatively unextendable commodity, either because of practical difficulties or because of the costs associated with doing so. It thus sees our epistemic condition as one in which we are allowed a relatively modest budget of experience or intuition to set the epistemic enterprise in motion and in which there is relatively little opportunity for causally prodding or massaging that modest initial budget of experience into a larger fund capable of meeting our epistemic needs. Therefore, we resort instead to *inference*, which, the classical view holds, offers us the epistemic benefits of extension of experience without the attendant costs and difficulties pertaining thereto.

Sometimes these "costs and difficulties" may take the form of sheer danger. While considering whether to dry my hair with my acetylene torch, it occurs to me that it may do to my head something like what it did to the pipe I cut with it last night. How do I decide what to do? I'm pretty sure that I do not want it to do to my head what it did to the pipe, but how do I find out if it will? I need to extend my knowledge in such a way as to decide whether the torch will cut my head like it cut the pipe. But I clearly do not want to do so by actually extending my experience in the appropriate way; that is, by actually trying out the torch on my head and observing what happens. The most elementary considerations of utility counsel against this, the disutility connected with failure being too high compared with the utility connected with success to make such trial-and-error experimentation rational. But how, then, do I extend my knowledge in the desired way *without* extending my experience?

The answer, roughly, is that I substitute a logically manipulated system of hypotheses for the physical act of applying the torch to my head. That is, I revert to a scheme of *representation* wherein the various states of my head and the torch are represented by propositions expressing those states, and the consequences of these states are then retrieved by retrieving the logical consequences of their representing propositions. For the *act* or *experience* of actually placing my head in the torch's flame, I thus substitute the *proposition* whose content is that I do so. And in the place of an experientially determined set of consequences of that act (i.e. the resulting burning sensation, the smell of burning hair and flesh, etc.), I substitute a set of propositions (representing those consequences) obtained by logical derivation from the proposition expressing the content of that act (together, typically, with certain auxiliary hypotheses representing the circumstances in which the act takes place and whatever natural laws may pertain thereto). I thus rely on a relatively painless *logical* manipulation of *representations* (propositions) rather than a potentially painful *experiential* manipulation of the corresponding *physical states* in order to determine what the consequences of drying my hair with the blowtorch would be. The happy outcome, as Popper once put it, is that I "permit my hypotheses to die in my stead."

An equally important, if less dramatic, illustration of the supposed need for the representational point of view is suggested by an empirico-constructional conception of geometrical thought like that found in Lorenzen (1984, 1985). On this conception, geometrical knowledge has its origins in a body of experiential constructional activities. For a variety of practical reasons, we are called upon to engage in such activities as the grinding of surfaces against one another to render them coplanar, the construction of planar figures using pencil, compass, and straightedge, the folding of planar objects to obtain other planar (or solid) objects, the rotation of these objects in various ways about the axes determined by such foldings, the construction of composite objects having a particular planar or solid character (e.g. that of being square) from component objects having another (e.g. that of being triangular), and so on.

The range and variation of such practical constructional activity, however, is strictly limited. Compass and straightedge can be practically managed only for planar objects of relatively small size; foldings and various other "reflection" operations must cope with such things as the tensile strength of the materials involved, the strength and accuracy of the folder, the length of her appendages, and so on. As a consequence of these limitations, we are not well situated to experientially determine what the result of folding, say, a one-block-on-a-side square of paper or a one-inch-on-a-side square of titanium will be. In short, our ability to actually extend our constructional activity to a wide variety of sizes and material-types of objects is strictly limited. To put it still another way, extension of knowledge over the full range of situations with respect to which we might desire such extension is simply too difficult to manage if we insist that it involve an extension of our actual constructional activity. Consequently, we seek a means of epistemically *projecting* our experience without actually *extending* it, so that our geometrical knowledge need not be bounded by the limitations of size, time, strength, etc. which limit our activities as actual line-drawers, paper-folders, planar-object-rotaters, etc.[6]

Extending this view beyond geometry to mathematics generally, we arrive at the classical viewpoint, which may be summarized as follows: mathematical knowledge may begin with a type of intuition or practice, but for a variety of reasons (having to do with the practical limitations concerning such things as our susceptibility to pain and the restrictedness of the time, effort, strength, material resources, etc. that we have to invest in such enterprises as the basic constructional activities of mathematics) this experience is insufficiently "plastic" to be practically extendable to the full variety of propositions over which we should like our knowledge to range; therefore, in place of the relatively impliant *practical* or *causal massaging* of mathematical *intuition*, we substitute a more pliant scheme of *logical* manipulations of its *contents*.

Thus it is that an experience comes to be represented by a proposition expressing its content. And, as with any good scheme of representation, one then uses more practicable (i.e. less dangerous, costly, etc.) operations on the *representens* to bring about the same basic epistemic effect as the less practicable operations on the *representanda*. This then is the general logic-intensive or representation-intensive conception of epistemic extension that we believe to have been the chief target of Brouwer's attack on classical mathematics.

IV. BROUWERIAN EPISTEMOLOGY

Brouwer offers the following hypothesis regarding the origins of the classical viewpoint:

> some very familiar regularities of outer or inner experience of time and space were postulated to be invariable, either exactly, or at any rate with any attainable degree of approximation. They were called axioms and put into language. Thereupon systems of more complicated properties were developed from the linguistic substratum of the axioms by means of reasoning guided by experience, but linguistically following and using the principles of classical logic.... [This viewpoint] considered logic as autonomous,

and mathematics as (if not existentially, yet functionally) dependent on logic.

(Brouwer 1951: 1)

He then goes on to identify as the (or at least a) fundamental mistake of this viewpoint the belief

in the possibility of extending one's knowledge of truth by the mental process of thinking, in particular thinking accompanied by linguistic operations independent of experience called "logical reasoning," which to a limited stock of "evidently" true assertions mainly founded on experience and sometimes called axioms, contrives to add an abundance of further truths

(Brouwer 1955: 113)

As an antidote to this basic miscalculation of the role of logical reasoning in the production of mathematical knowledge, Brouwer offered his so-called First Act of Intuitionism,

completely separating mathematics from mathematical language and hence from the phenomena of language described by theoretical logic, recognizing that intuitionist mathematics is an essentially languageless activity of the mind having its origin in the perception of a move of time.

(Brouwer 1951: 4)

According to Brouwer, mathematics is essentially a form of introspective constructional *activity* or *experience* whose growth or development thus cannot proceed via the logical extrapolation of its content (as classical epistemology maintains), but rather only by its *phenomenological* or *experiential* development – that is, its extension into *further experience* of the same epistemic kind. The logical extrapolation of content – that is, logical inference – can never, Brouwer says, "deduce a *mathematical* state of things" (cf. 1954: 524; emphasis added). In genuine mathematics, theorems are proved "exclusively by means of introspective construction" (cf. 1948: 488). Logical laws are not "directives for acts of mathematical construction" (cf. 1907: 79), but rather derive from regularities in the language (possibly mental) used to express or represent such constructions. And while the regularities of a given such scheme of representation might prove useful in our attempts to remember genuine mathematical experiences, and to communicate them to others, they must not be confused with or equated to means of actually extending that experience (cf. 1907: 79; 1908: 108; 1955: 551–2). Indeed, we must bear in mind the fact that, even judged solely as instruments for memorization and communication, such schemes for representing experience are subject to limitations of exactitude and correctness (cf. 1951: 5).

Mathematical knowledge is thus essentially a form of constructional activity, with the consequence that extension of that knowledge must take the form of extension of that activity, rather than a mere, actionally disembodied, logical extrapolation of its contents. This, at any rate, is the anti-classical kernel of Brouwerian epistemology which is of principal concern to us in this essay.

Brouwer's central thesis, then, is the general and sweeping one asserting the *experience-intensive* – and denying the *logic-intensive* – character of mathematical knowledge and its growth: mathematical knowledge is a form of experience or activity, and growth of mathematical knowledge therefore requires growth of that activity. Thus, if mathematical knowledge of a proposition **p** is to be extended to mathematical knowledge of a proposition **q**, the experience or activity whose content is **p** must be *transformed into* an experience or activity whose content is **q**. In contradistinction to the classical model of epistemic growth, then, Brouwerian epistemology does not present the prover as reflecting on contents, generating new from old by this logical reflection, and thence transferring the warrant for the old to the new (by appeal to the warrant-preservingness of the modes of contentual analysis employed). Rather – and it is hard to overemphasize the importance of this difference to the present discussion – the mathematician transforms old proof-experiences or proof-activities into new ones and thus witnesses the extension of her knowledge to new propositions when such a proposition emerges as the content of the newly created proof-experience. What is thus crucial and basic is the creation of a new proof-experience. For once such experience exists, knowledge may be extended to whatever its content is. What logical relation the content of this newly created experience might bear to that of the old is a matter of secondary concern. For knowledge-extension proceeds not by the logical extraction of new propositions from ones already known, but rather by the phenomenological transformation of one proof-experience into another – the next content emerging as *the content of* the new experience produced by this transformation. Mathematical inference or proof thus follows the path of the possibilities relating mathematical activities, rather than the chain of connections determined by some logico-linguistic analysis of the (propositional) contents of such activities, as classical epistemology maintains.

This, then, is the substance of Brouwer's *first* – in order of basicness and importance, if not of recognition – critique of classical mathematics. It faults classical epistemology not for its particular choice of logical principles to serve as means of extending mathematical knowledge, but rather for the fact that it accords such epistemological power to *any* set of purely logical principles. For the Brouwerian, a proof is more (and also less) than just a series of epistemic attitude-takings whose contents are logically related. It follows instead an ordering of activities where what might be called the "actional accessibility" of one constructional activity from another is more important than the logical accessibility of the content of the one from the content of the other.[7]

This general insistence on the part of Brouwer to distinguish between the logical extrapolation of mathematical knowledge, on the one hand, and the genuine extension of mathematical knowledge on the other, should not, however, be taken to imply that he denied any and all epistemic significance to logical inference. For he seems to have granted to logical extrapolation a certain limited role as an *instrumental device* (founded on the manipulation of a scheme of representation for proof-experiences which represents them by means of their propositional contents) for identifying, remembering and communicating propositions for which an intuitionistic proof-experience might be found.[8]

Determining by means of a logical "calculation" that a given proposition **p** can be given an intuitionistic proof is not, of course, epistemically equivalent to either giving or being in a position to give an intuitionistic proof for **p**. Nor is a logical derivation which determines that an intuitionistic proof of **q** can be obtained from an intuitionistic proof of **p** the same as either transforming or being in a position to transform a proof-activity for **p** into a proof-activity for **q**. Yet despite the fact that a logical "calculation" that **p** is provable is epistemically inferior to either having or having the practical ability to produce a proof for **p**, it does not follow that it is of no epistemic value whatever. It can have value – as a device for determining where to invest one's proof-seeking efforts – can, that is, to the extent that it is accurate.[9]

It is this matter of accuracy that stands behind Brouwer's *second* (though better known) critique of classical mathematics; namely, the critique of the law of excluded middle and allied principles of classical logic.

> the function of the logical principles is not to guide arguments concerning experiences subtended by mathematical systems, but to describe regularities which are subsequently observed in the language of the system ...
> Thus there remains only the more special question: "Is it allowed, in purely mathematical constructions and transformations, to neglect for some time the idea of the mathematical system under construction and to operate in the corresponding linguistic structure, following the principles of *syllogism*, of *contradiction* and of *tertium exclusum*, and can we then have confidence that each part of the argument can be justified by recalling to the mind the corresponding mathematical construction?"
> Here it will be shown that this confidence is well-founded for the first two principles, but not for the third.
> (Brouwer 1908: 108–9)[10]

Brouwer's critique of the law of excluded middle therefore had the status not of an argument designed to show that though other logical principles might play a significant role in the construction of a proof it (i.e. excluded middle) cannot. Rather, it had the status of a critique of a "calculating" device; a device which, even if perfect, could play no serious role in the giving of genuine proofs, but rather could serve only as a means of identifying those propositions that might be given a proof. Brouwer's criticism is that, used as (part of) a device for locating those propositions capable of being the contents of an intuitionistic proof-experience, excluded middle would lead to the identification of certain propositions as having this trait when in fact they do not. Therefore, it is unsatisfactory as (part of) an instrumental device for "calculating" which propositions have the potential to become contents of intuitionistic proof-experiences.

In addition to the inaccuracy borne of this *unsoundness*, there is another respect in which classical logic is inaccurate. This inaccuracy stems from its *incompleteness* as a device for locating those propositions that are capable of receiving a proof. Brouwer argued this point vigorously, and developed a battery of results from analysis which he took as illustrating it (cf. Brouwer 1923,

1949a,b). Among these, perhaps the most famous is his proof of the Continuity Principle; that is, the theorem stating that every total real-valued function on the closed unit interval is uniformly continuous (cf. Brouwer 1923: 248).

Brouwer could thus sum up his criticism of classical logic as an instrument for determining which propositions are capable of being the contents of an intuitionistic proof-experience by saying that "there are intuitionist structures which cannot be fitted into any classical logical frame, and there are classical arguments not applying to any introspective image" (cf. 1948: 489). The first part of this claim emphasizes the inaccuracy borne of the *incompleteness* of the classical instrument, while the second emphasizes that which results from its *unsoundness*. If the principles of classical logic were to be amended in such a way as to eliminate these deficiencies of incompleteness and unsoundness, then one would have an apt logical instrument; that is, an accurate device for determining which propositions are potential contents for intuitionistic proof-experiences. However, such a device could still serve only to *identify* those propositons that are capable of intuitionistic justification – which is a very different thing from (and epistemically inferior to) actually supplying such justification.

Such, at any rate, is our understanding of the Brouwerian standpoint, which is strikingly at odds with the usual version of intuitionism presented in the literature. On the usual version, the critique of excluded middle is presented as the centerpiece of the intuitionist's concerns and the crux of his criticism of classical mathematics. Our view differs from this in two ways. First it suggests that the question "Which logic is the logic of mathematics?" (and particularly the subquestion "Does the law of excluded middle belong to the logic of mathematics?") is of secondary importance. The more fundamental question is "What role does *any* logic (including the "right" one) have to play in the construction of intuitionistic proofs?" Judged from this vantage, the question "Which logic is the logic of mathematics?" can only be regarded as misleading.

The second respect in which our view differs from the usual one is in its deflation of the significance of the critique of excluded middle – even with respect to the role that it plays in the criticism of classical logic as a locative/mnemonic device. On the view presented here, that critique is to be seen as but one part of a larger two-part critique that is concerned not only with the soundness of the classical locative/mnemonic device, but also with its completeness. Basically, the critique of excluded middle is a critique of soundness and makes little if any contribution to the assessment of the completeness question, despite the fact that the latter question is just as important to the accuracy of a locative/mnemonic device as the soundness question.

It may even be that the importance of the critique of excluded middle should be deflated still further. For of the two parts of the accuracy question, the part to which it contributes (namely the soundness question) may be of less overall significance to Brouwerian epistemology. To understand why this is so, we must hearken back to what we identified in the last section as the basic motivation of classical epistemology.

That motivation, it will be remembered, had its basis in the conviction that intuition or experience is a relatively scarce epistemic commodity – that it is not readily accessible in sufficient quantities to beings subject to the practical limitations (e.g. of strength, size, sensitivity to heat, flammability, etc.) that we are. Therefore, the classicist seeks a way of liberating knowledge from its meager intuitional or experiential origins. His answer is the logic-intensive or representation-intensive stratagem. On this stratagem, warrant is identified with some property (e.g. certainty, or certainty plus such things as *a priori* status) that is relatively insensitive to the fine points of the cognitive mode of a warrant and focuses more on its content. As a result, it (i.e. warrant) becomes the sort of thing that can be passed on by techniques of inference that preserve relatively few of the details of the cognitive mode under which the premises of the inference are presented as warranted.

This way of motivating classical epistemology presents a challenge to the Brouwerian. For if intuitional knowledge really is as rare and hard to obtain as the classicist says, then how can the Brouwerian hope to build a thriving epistemic enterprise while at the same time repudiating the classicist's model of epistemic growth? We believe that Brouwer's critique of the incompleteness of classical reasoning can be seen as speaking to this concern. In arguing for the incompleteness of the classical logico-linguistic method, what is being brought out is that not all the liabilities for epistemic growth lie on the side of the intuitionist. The classicist too has liabilities, there being things that he cannot prove that his intuitionist counterpart *can.* The result is that the motivation of the classicist's logic-intensive approach to epistemic expansion is to some extent blunted, since it is no longer clear that epistemic expansion via logical manipulation of the content of knowledge has greater productive potential than epistemic expansion via extension of intuition.[11]

The account of Brouwerian epistemology as sketched up to this point emphasizes the effects brought about by the prominence it gives to occurrence in the *experiential mode* as an important trait of mathematical knowledge. That emphasis may appear to be lacking in motivation, however. To address that need, we must now sound some deeper themes of Brouwerian epistemology.

Let us begin by recalling an oft-recited tenet of intuitionism that forms the cornerstone of Brouwer's outlook. This tenet is the deceptively simple, though in truth quite radical, idea that mathematics, in its essence, is a form of mental *activity*. We propose to take this emphasis on the *actional* or *practical* character of mathematics seriously, and thus to investigate the possibility of treating Brouwerian epistemology as based on a *practical* rather than a *theoretical* conception of mathematical knowledge.

On this way of looking at it, mathematics is a body of actions or capacities for action, rather than a body of truths (i.e. a *science*, in the traditional sense). Similarly, mathematical knowledge is a type of *practically-knowing-how* to perform certain actions, rather than a rational reflection on various propositions and a subsequent *intellectual-recognition-that* they are true. We also intend to take this distinction between practical and theoretical knowledge as ultimate. That is, we propose to interpret it in such a way as to imply the inconvertibility, at least to the point of epistemological equivalence, of the former sort of knowledge into the latter.[12,13] The mental activities of the intuitionist, like the attitude-takings or "acceptances" of conventional epistemology, may be thought of as having propositional contents. But, in being *lived* or *experienced*, those contents are epistemically "registered" in a way that is not reducible – at least not without epistemic loss – to any kind of purely intellectual grasp of them. The emphasis on this "livedness" or "experienced-ness" is a way of expressing the practical character of the knowledge involved. We "live" our activities. Thus, since mathematical knowledge is ultimately an activity or capacity for activity, it will ultimately manifest itself through our experiencing of our practical lives.

(N.B. In addition to this, the emphasis on experience may also be partly an attempt to express the idea that there is somehow something of greater value in a kind of knowledge that brings with it a capacity to *do* something than in a kind of knowledge which consists solely in an intellectual "acknowledgement" or "acceptance" of a proposition. Genuine knowledge – so the idea would go – enlivens and enables. It moves to action. It is more than just the doffing of one's intellectual hat to a proposition. Practical knowledge therefore penetrates to a level of our cognitive being to which theoretical or purely intellectual knowledge typically does not.[14])

In an epistemology thus dominated by a practical conception of knowledge, it should come as no surprise that such accoutrements of the theoretical or scientific conception of knowledge as the use of logical inference and the axiomatic method are devaluated, and concern for the convertibility (or, to use the term that we have been using, the "transformability") of one activity or practical capacity into another is put in their place. Thus, on the epistemology being sketched here, an area of mathematical thought (the correlate of a *mathematical theory* under the traditional conception) is to be thought of as a body of actions organized by a scheme of actional connections reflecting some sort of practical disposition to pass from one *act* to another, rather than a body of truths organized by a network of logical relations. Likewise, in place of a plan for epistemic growth which sees it as a march from one intellectual "acceptance" to another via the steady logical exploitation of the propositions thus accepted, towards a goal of "complete" acceptance (that is, acceptance of the complete set of truths pertaining to the subject-matter of the science in question), there is a course of practical development which is seen as consisting in the practical transformation of one act into another in such a way as to bring one's overall mathematical activity into closer conformity to a network or "stream" of actions which is taken to represent the ideal of an abundant mathematical *life*.

Both the goal and procedure of epistemic development thus change when one moves from a theoretical to a practical conception of mathematical knowledge. In place of a goal of "complete" theoretical knowledge, we have the ideal of an abundant practical *life*, reckoned not (or at least not primarily) in terms of the logical properties (e.g. consistency, completeness) of the set of propositions known, but rather in terms of the practical power which its activities represent. And, in place of epistemic extension of the domain of our intellectual "acceptances" from one proposition to another via logical inference, there is the extension of our practical capacities which is based on the acquisition and realization of dispositions which link one mathematical activity to another. Thus, our "local" or individual proof-activities come to be bound together into a global whole (a *life*) by a scheme of relations which are not constituted by the logical relations which prevail among their propositional results, but rather by their actional or behavioral affinities to one another. Different local proof-activities are thus to be seen as exhibiting not only a logical relationship between their contents, but also a global relationship of "fit" or "continuity" which reflects a *practical* disposition to move from performance of one act to performance of another in such a way as to draw nearer to the ideal of an abundant mathematical *life* – defined too in terms of *practical* accomplishments and capacities. Correct global orientation at a given locale (i.e. for a given local proof-activity) is thus a matter of that local activity's being dispositionally related to other local proof-activities in such a way that, allowed to develop in a natural way, they would grow into a body of proof-activities having the sort of practical potency that is seen as being constitutive of mathematical maturity.

One feels, of course, the need for some description of the above-mentioned dispositions which characterizes them as something other than a set of dispositions which, allowed to develop naturally, would lead to global configurations of proof-activities (mathematical lives) having the desired sort of potency and integrity. One needs a description of them which reveals why they should be expected to lead to a body of proofs having the desired global integrity. Perhaps Brouwer's singling out of the unfolding of the bare notion of two-ity in the mind at perfect rest, with no "sinful" designs on the conquest of nature, and no "cunning" or even "playful" attempts to manipulate the stream of inner experience, can be seen as bearing on such a concern: those proof-activities which are dispositionally related to other proof-activities in such a way as to grow into the right sort of global practice are those of the mind at perfect (causal-manipulatory) rest, with no designs on causal dominion over nature or even over one's own stream of inner mathematical experience.

The practical-knowledge model just sketched is but one attempt to flesh out, in a Brouwerian manner, the central theme of our argument: namely, that, as Poincaré pointed out, mathematical reasoning appears to differ fundamentally from logical reasoning, and that in order to account for this difference one must seemingly reject the classical logic-intensive epistemology for mathematics. It is not, however, the only way of proceeding, as we shall now briefly attempt to indicate by sketching a Brouwerian theoretical-knowledge model of mathematical knowledge.

On this model, the emphasis on the "experienced-ness" or "lived-ness" as the distinctive cognitive mode of mathematical knowledge, which figures so prominently in the practical-knowledge model, is replaced by an emphasis on the "locality" of one's theoretical knowledge. Basically, the idea is this: one's knowledge of a mathematical truth **p** is *mathematical* to the extent that it is based on a "local" familiarity (in the sense discussed above in Section II) with the mathematical subject(s) to which **p** belongs. This emphasis on the "local" character of mathematical knowledge seems to be but another way of putting Brouwer's point concerning the "autonomy" of mathematical thought, which was that we ought to be careful to distinguish the connections between propositions which arise for the *linguistic representation of* mathematical reasoning from the connections between propositions which characterize that reasoning itself, and therefore not attribute to mathematical reasoning "regularities in the language which accompany it" (cf. Brouwer 1955: 551–2). "Regularities of language" are to be expected to be of a global character, since languages are intended to constitute global schemes of representation; that is, schemes of representation designed not with the representation of some one body of thought in mind, but rather of *all* bodies of thought generally. It is therefore not to be wondered that the linguistic representation of mathematical thought should induce a global logical structure on its theorems. Nor is any harm done by this so long as it is remembered that this induced logical structure (which truly deserves to be called logical because of its global character) is to be taken as a structure imposed by the representing device, and not as the structure of the thought being represented. As Brouwer said, "Mathematical language, in particular logic, can never by itself ... deduce a mathematical state of things" (cf. 1954: 524).

Whatever structure is exhibited by genuinely mathematical reasonings thus appears to be of a more "local" character, determined by the subject of the constructional thinking in question. Such constructional thinking may, of course, be divisible into steps or parts in a variety of ways. But the question is which decompositions into steps actually correspond to the step-structure exhibited by a genuinely mathematical piece of reasoning, and which merely represent different not-genuinely-mathematical ways of systematically decomposing the same complex thoughts. Brouwer's emphasis on the "autonomy" of mathematics with respect to logic suggests that, though logical structures may sometimes be superimposed on complex mathematical reasonings, they just represent a sort of "tacking together" of mathematical affirmations (cf 1933: 443) that is "co-extensional," as it were, with the genuine mathematical reasoning on which it is superimposed. They do not, however, generally reflect the structure of that reasoning *considered as genuinely mathematical reasoning.*

There is thus an epistemological basis for a Brouwerian repudiation of classical logic even if one inclines to a theoretical rather than a practical conception of mathematical knowledge, and one also wishes to avoid any appeal to private phenomenological characteristics of such theoretically conceived knowledge. And the key element of that basis is nothing other than Poincaré's point concerning the "locality" of genuinely mathematical knowledge – hence our emphasis on Poincaré's point as furnishing a basis for Brouwerian epistemology.

V. INTUITIONISTIC LOGIC

We would like to close by making a few remarks about an implication of our position that seems likely to puzzle; namely, the great disparity between what we have portrayed as the Brouwerian intuitionist's attitude towards logic, on the one hand, and that of the present-day intuitionist, on the other. The latter extends a much greater role to the use of logical inference in mathematical reasoning than does the former. In this the final section of the essay, we shall briefly consider some of the assumptions made by the present-day view in order to determine what would be required for their justification. Particularly, we shall center our attention on those assumptions concerning the manipulability of mental mathematical constructions that are needed for the defense of intuitionist logic, and consider their plausibility as structural characteristics of a domain of mental constructions of the practical or theoretical varieties described in the last section. In the end, our finding is negative: the assumptions concerning the manipulability of mental mathematical constructions that seem to be needed for the defense of intuitionist logic are not plausible when considered as features of the practical or theoretical knowledge described in the last section.

To begin our discussion, let us consult the *locus classicus* of present-day intuitionism – namely, Heyting (1956), where the standard conception of mathematical constructions as proofs is introduced as follows:

> a mathematical proposition **p** always demands a mathematical construction with certain given properties; it can be asserted as soon as such a construction has been carried out. We say in this case that the construction *proves* the proposition **p** and call it a *proof* of **p**.
>
> (Heyting 1956: 102)[15]

Having thus characterized the basic condition of proof as possession of a construction, Heyting then went on to give a more detailed description of the specific assertion-conditions pertaining to the different kinds of compound propositions that can be formed by applying the various logical operations to simpler propositions. Naturally, this description was guided by a certain view of the dynamics of construction-possession; that is, a view of the general laws according to which possession of a given set of constructions induces possession of others. Thus, an inductive scheme stating how possession of constructions for logically compound propositions is related to possession of constructions for simpler propositions is given.[16] Using "$\pi(\chi, A)$" to stand for "χ is a construction which proves A," that scheme is something like the following:[17]

(a) $\pi(\chi, A \& B)$ iff $\chi = \langle \chi_A, \chi_B \rangle$ and $\pi(\chi_A, A)$ and $\pi(\chi_B, B)$.
(b) $\pi(\chi, A \vee B)$ iff $\chi = \langle \chi_A, \chi_B \rangle$ and $\pi(\chi_A, A)$ or $\pi(\chi_B, B)$.
(c) $\pi(\chi, A \rightarrow B)$ iff for all constructions κ, if $\pi(\kappa, A)$, then $\pi(\chi(\kappa), B)$.
(d) $\pi(\chi, \neg A)$ iff for all constructions κ, if $\pi(\kappa, A)$, then $\pi(\chi(\kappa), \bot)$, where $\bot$ is some agreed-upon intuitionistically refutable proposition.
(e) $\pi(\chi, \exists x Ax)$ iff there is a (term-) construction τ and a (proof-) construction κ, such that $\chi = \langle \kappa, \tau \rangle$ and $\pi(\chi, A(\tau))$.
(f) $\pi(\chi, \forall x Ax)$ iff for each number n, $\pi(\chi(n), A(n))$, where "n" is the standard numerical term corresponding to n.

Basic to these conditions, which purport to state the laws that regulate the interaction between construction-possession for simple propositions and construction-possession for compound propositions, are a set of structural constraints induced upon the domain of intuitionistic constructions by the following general conditions: (i) that any two constructions can be "paired" in order to yield a new construction, (ii) that constructions constituting proofs for compound propositions can be decomposed in such a way as to yield constructions for select simpler propositions, and (iii) that there are constructions that can be applied to constructions in order to yield constructions.

The need for the first condition is illustrated by clause (a), taken in the right-to-left direction (which sanctions the introduction rule for the & operator). There, the possession of separate constructions for A and B is parlayed into possession of a construction for A & B by "pairing" the individual constructions for A and B into a single compound construction which then serves as a construction for A & B.

The need for the second of the above-mentioned structural conditions may also be illustrated by reference to clause (a), this time, though, taken in the left-to-right direction (which sanctions the elimination rule for &). There, one begins with a single construction for the compound proposition A & B and decomposes it into component constructions which are constructions for A and B. Thus, decomposition of constructions for compound propositions into constructions for simpler propositions (in the case of &-compounds, actual components of the compound proposition for which it is assumed that one has a construction) is assumed to characterize the realm of intuitionistic constructions.

The need for the third of the above-mentioned structural conditions is well illustrated by the clause for the conditional operator. There it is assumed that the construction χ is the sort of thing that can itself be *applied to* constructions for A in order to obtain constructions for B. Thus the domain of intuitionistic constructions is taken to contain certain constructions which themselves are the sorts of things that operate on constructions with the result of producing other constructions.

The effect of conditions (i)–(iii) is thus that of creating a kind of "algebra" for the domain of intuitionistic constructions. The task facing us is that of determining how plausible the "algebra" thus conjectured is when it is taken as an algebra for the constructions of the practical-knowledge or theoretical-knowledge models of the last section.

For the sake of concreteness, we shall present our argument with specific references to the operation of logical conjunction. Arguments similar to the one we are about to present could be given with respect to the other operations as well, but the problems which we are concerned to bring to light arise in their clearest and simplest form in the case of conjunction. Also, the focus on conjunction seems to be strategically well taken since present-day intuitionists appear to regard it as the least problematic of all the logical operations. Thus, any problems detected in connection with it are likely to point to basic difficulties. Let us consider, then, the claim that a proof of A & B may be obtained from a proof of A and a proof of B by a supposed operation of "proof-pairing."

It is easy to miss the problem presented by such a claim. For it is tempting to mistake what is, in actuality, a claim concerning the behavior of a *natural* kind of mental activity (namely the intuitionist's mental mathematical constructions) for a claim concerning some sort of purely logical or conceptual possibility. In other words, it is tempting to think that the claim in question can be defended by simply pointing to the conceptual possibility of binding a proof of A and a proof of B into some sort of unit, or considering them "together" in thought.

On the practical-knowledge model sketched in the last section, the mental mathematical constructions of the intuitionist form a *natural* kind; that is, a kind of activity whose laws of combination are governed not by mere conceivability, but rather by the laws governing the natural kind of the activity involved. The mere fact (if it is a fact) that we can always conceive of a construction of A and a construction of B being bound together into a single construction of A & B is not enough to show that the natural kind of mental mathematical construction also operates in this way. We can perform logical operations on the contents of our constructions, but they may run transverse to rather than being coincident with those laws governing the stream of constructional activity as a natural epistemic kind. "Pairing" is thus a mere tag for an undescribed and rather dubious (because it runs contrary to the generally observed autonomy of constructional activity from the machinations of logic) mental operation that is supposed to allow us to take any two separate constructional acts and turn them into a single (complex) act whose content is the conjunction of the contents of the separate experiences.

Similar remarks apply to the theoretical-knowledge conception.

The fact that one has a local proof of A and a local proof of B does not imply that one either does or can have a local proof of A & B (though this might of course hold for certain particular A and B, and even for certain limited *kinds* of A and B). This is particularly clear if A and B are drawn from different local settings, but it is to be generally expected even for A and B drawn from the same local setting, since the reasoning according to which the local proof of A and the local proof of B are to be bound together is not local but rather global reasoning. Hence, it is not by local insight that the two proofs are bound together, and this may be enough to deprive the compound proof of status as local reasoning.

The attractiveness of conjunction-introduction and the other rules of "intuitionist logic" may be due in large part to the fact that, since Heyting's original formalization of intuitionist logic, most of the work done on the subject has been of a technical rather than a philosophical nature and that this work has yielded clear, precise, and effectively executable *syntactical counterparts* for the mentalistic operations on proof-constructions (namely proof-pairing, proof-decomposition, and construction-application) that would be required by a genuine logic of intuitionistic reasoning. Being effectively executable, these syntactical operations are, of course, intuitionstically acceptable, and so it may be that this has led some to lose sight of the fact that, though they are of an intuitionistically acceptable character, they are nonetheless procedures not for operating on intuitionistic proofs *per se*, but only on certain of their *syntactical representations*.

Nor do such syntactical or formal operations on proof-representations *suggest* any clear parallel operation at the level of genuine mentalistic proof. This can, perhaps, be made clear by considering the syntactical counterpart of proof-pairing; namely, syntactical concatenation. Concatenation is an operation that calls for the sequential arrangement of concretia in space–time. However, one certainly cannot produce a "compound" mental proof by laying two "smaller" proofs end-to-end. Indeed, it is unclear what it would *mean* to lay mentalia end-to-end. Even adding the (by no means obvious) supposition that mental states are to be regarded as concretia of some sort (e.g. brain states), what would syntactical concatenation suggest as a parallel at the level of mentalia? Spatial contiguity of brain states can hardly be expected to be what "concatenation" of mental proofs would come to – nor can temporal contiguity, since there are many temporally contiguous mental states that cannot be made into any kind of meaningful mental unity at all (e.g. a state corresponding to a mental mathematical construction followed by a state corresponding to my being startled by a loud noise).

It therefore seems clear that the syntactical operation justifying conjunction-introduction in *formalized intuitionist logic* cannot be taken as justifying conjunction-introduction as a rule of *genuine intuitionistic proof*, since though "concatenation" may be clear as an operation on syntactical entities, it gives no indication of what its counterpart at the level of mental proof might be.[18]

The above remarks are directed at a conception of intuitionistic conjunction which sees it as based on the philosophy of pairing proof-constructions of A and B to form a new proof-construction whose content is A & B. There is, however an alternative way of thinking of intuitionistic conjunction (cf. Dummett 1977: 12; Martin-Löf 1983: *passim*). On this view, a proof of A & B is not to be seen as some third, compound entity (χ_A, χ_B), distinct from both χ_A and χ_B (the proofs for A and B, respectively), yet formed by somehow combining them into a single construction whose content is A & B. Rather, it is to possess the separate, individual constructions χ_A and χ_B in a certain way; to have them, as it were, "simultaneously."

This way of thinking of conjunction avoids the need to show that there is a mental pairing operation that preserves local character (in the case of the theoretical-knowledge approach to intuitionist epistemology) or that preserves the natural kind of intuitionist constructional activity (in the case of the practical-knowledge conception) when it is used to join separate proofs to form a compound proof of their conjoined contents. For it allows there to be a construction of the compound proposition A & B without having a single construction whose content is the conjunction of the contents of a construction of A and a construction of B. On this account, possession of separate constructions for A and B is all that is required, provided, of course, that that "possession" is understood as providing equal access to both constructions at a given time.

Such an account may be adequate so far as conjunction-elimination is concerned since all it requires is that a proof of A & B be the sort of thing that affords one access to both a proof of A and a proof of B – and "simultaneous" possession of proofs for A and B is exactly that sort of thing. Conjunction-introduction, however, is another story. There, it seems that a proof of A & B is to be more than a mere provider of equal access to individual proofs of A and B. It is to be something that moves from simultaneous possession of separate proofs for A and B to a proof which synthesizes their respective contents. Indeed, it is that alone which qualifies it as a candidate for genuine inference. For genuine inference demands that there be some "movement" (i.e. some change of mental state) in going from the premises of an inference to is conclusion. Hence, whatever state it is that is taken to constitute an epistemic grasp of the conclusion must be different from that which is taken to constitute a grasp of the premises.

This condition is satisfied for conjunction-elimination even when possession of a proof for A & B is taken to consist merely in what we have been calling "simultaneous" possession of proofs for A and B. For the conclusion of an inference by conjunction-elimination demands only a grasp of a proof for A or a proof for B, whereas the premise demands a simultaneous grasp of both. But the same is not true of conjunction-introduction. There, the premises already demand simultaneous possession of a proof for A and a proof for B rather than mere separate possession of a proof for A and a proof for B, since one needs to hold the proofs together as a unit of some sort in order to ascend to the conclusion. Having a proof of A and a proof of B, but not being able to bring them together, would not allow one to do anything other than offer a proof of A and, separately, a proof of B. Thus, there is a difference between holding a proof of A and a proof of B separately and holding them together as a unit of some sort (i.e. holding them "simultaneously"), and it is the latter which is required by the premises of a conjunction-introduction.

This being so, and it also being the case that genuine inference demands "movement" of the sort described earlier, we may infer that the conclusion of an inference by conjunction-introduction demands more than just the simultaneous possession of proofs of A and B – at least to the extent that conjunction-introduction is to constitute a form of genuine inference. The extra that is needed is a synthesizing of the contents of the proofs of A and B that introduce the premises. The intuitionist's mental mathematical proof-constructions (as opposed to object- or term-constructions) are *intentional* states; that is, mental states that are *about* (constructed) objects, and which thus have a propositional content.[19] Hence, ordinarily, or perhaps we should say *canoni-*

cally, to have a mental mathematical (proof-) construction for a proposition σ is to be in an intentional state μ having σ as its content, and occurring in a certain mode M_μ which reflects its "local" origins (in the case of the theoretical-knowledge model), or its membership in the natural kind of intuitionistic constructional activity (in the case of the practical-knowledge model). Derivatively, or noncanonically, to have a mental mathematical construction for σ is to be in a mental state μ′, the being in of which facilitates access to (i.e. puts one – at least ideally – in a position to produce) a canonical mental mathematical construction μ whose mode is M_μ and whose content is σ.

In either case, however, having a mental mathematical proof-construction of A & B requires either being in or having access to an intentional state whose content is A & B. This makes having a proof-construction for A & B different from merely having a single means of producing both an intentional state whose content is A and an intentional state whose content is B, which is what "simultaneous" possession of a proof-construction for A and a proof-construction for B comes to. Hence, to be a genuine form of inference, conjunction-introduction must be seen as moving from "simultaneous" possession of proof-constructions for A and B to a proof-construction having A & B as its content. This being so, the view of intuitionistic conjunction which maintains that to have a proof-construction of A & B is just to have a proof-construction of A and a proof-construction of B cannot be accepted.

We have thus examined the two contemporary conceptions of intuitionistic conjunction and found them both lacking any plausible account of how conjunctive inference might be blended into the intuitionist's mental procedures. And what has been said of conjunction would appear to apply to other parts of intuitionist logic as well. We conclude, therefore, that the role of intuitionist logic in intuitionist mathematical reasoning is quite suspect, and that the "algebra" of constructions upon which that logic is supposed to be based is rather dubious as a description of the ways in which complex constructions are formed intuitionistically.

VI. CONCLUSION

The account of Brouwerian intuitionism sketched in this paper is one which seeks to give it a new emphasis. Instead of the privacy or interiority of Brouwer's "autonomic interior constructional activity" (cf. Brouwer 1955: 551), we have stressed its *autonomy* and claim to find in this autonomy a basis for his radical rejection of logic which avoids the solipsistic pitfalls of an approach based on interiority.[20] We have, moreover, presented two different ways that one might go about developing this emphasis on autonomy: one a practical-knowledge model which stresses the autonomy of structure which proof-construction, as a natural kind of activity with its own natural kind of "movement," enjoys; the other a theoretical-knowledge model where autonomy derives from the fact that what powers the flow of knowledge is local insight rather than global inferential possibilities. Both provide bases for minimizing the role of logical inference in mathematical reasoning.

Both, too, are arrived at as transcendentally deduced hypotheses designed to explain a prior "datum" of mathematical epistemology − namely, the observation that there is a seemingly important difference between the epistemic condition of the genuinely mathematical reasoner and Poincaré's "logician" (i.e. one who arrives at her conclusions via a series of logical machinations). This observation does not so much call into question the feasibility or plausibility as the very *desirability* of an intuitionistic logic. For it suggests that, should the intuitionistic logician's advice be followed, the plausibility of intuitionist epistemology would be put in jeopardy, since it would then lose the ability to account for the difference between the epistemic quality of reasoning that is based on a genuine mathematical knowledge and inference that is not.

We believe that Poincaré's observation has been unjustifiedly ignored as a desideratum of mathematical epistemology. Likewise, we judge Brouwerian intuitionism, with its emphasis on privacy dropped and refocused on the autonomy of mathematical reasoning (with the result that it is able to respond to Poincaré's Concern), to have been underestimated as a mathematical epistemology. Our hope is to have taken some steps towards correcting these oversights, and to have revealed some of the interesting and challenging features of the views of Brouwer and Poincaré which, for the most part, have escaped the notice of contemporary philosophers of mathematics.

NOTES

*First published in *Mind*, vol. 99, July 1990, © Oxford University Press 1990. Reprinted with the kind permission of the editor. The author wishes to thank audiences at the University of Wisconsin-Madison, Calvin College, Oxford University, the University of Konstanz, and the University of Notre Dame for useful discussions. Individuals whose discussion has been helpful include Michael Byrd, Kosta Došen, Daniel Isaacson, Friedrich Kambartel, Georg Kreisel, Philip Quinn, Stewart Shapiro, Pirmin Stekeler, and G.H. von Wright. Special thanks are also due to the Philosophy Department at the University of Konstanz for its fine hospitality during the 1987-8 academic year, and to the Alexander Von Humboldt Foundation for its generous financial support.

1 Poincaré's repudiation of logical inference as belonging to genuine mathematical reasoning, of course, necessitates the development of a notion of rigor that is different from the usual logical one (according to which a proof is rigorous only if each of its steps of inference is purely logical). Poincaré himself did not say a great deal about this. Still, certain of his remarks suggest a radically new conception of rigor, one which sees rigor as consisting in the elimination of gaps in our *mathematical understanding* rather than logical gaps. On this model, an inference is rigorous only if we have a truly *mathematical* (hence, topic-specific) insight into why the premise's being true insures the truth of the conclusion. The radicality of this suggestion can be seen from the fact that it not only allows nonlogical inferences to be rigorous, but also implies that purely logical inferences (based on topic-neutral knowledge) are not rigorous!

It should also be noticed that Poincaré's point, though closely related to a famous point of Kant's, is nonetheless different. Kant maintained that some analytic inferences can be epistemically fruitful; if their conclusion is buried deeply enough in their premises, then digging them out can bring new knowledge. Moreover, he appears to have allowed that this hold for extensions of *mathematical* knowledge. Poincaré, on the other hand, though he may have allowed that analytic inference could sometimes produce new knowledge of some types, nonetheless explicitly denied that this is so for mathematical knowledge.

2 It may be helpful to say just a few words comparing the positions of Brouwer and Poincaré on this point. Both describe the special mode characterizing mathematical knowledge as knowledge by "intuition." However, they do not mean the same thing by that. For Poincaré, as was mentioned above, intuition is taken to be constituted by some sort of integrated knowledge – ultimately theoretical rather than practical in character – which enables the mathematical knower to see how a certain proof or theorem relates to other proofs and theorems, and how, in thus relating, it contributes towards the goals of some larger inquiry to which they all belong, and for the sake of which they are pursued. For Brouwer, on the other hand, the epistemically salient

and distinguishing feature of intuition is that it is a type of knowledge borne of experience of an ultimately practical nature, and thus basically incapable of extension by logical inference. More on this later.

3 "Pre-intuitionist" was Brouwer's term for the philosophical views of the so-called "French School," whose chief figures were Borel, Lebesgue, and Poincaré. The "intuitionist" part draws its justification from the fact that, according to Brouwer, these thinkers regarded the objectivity and exactness of certain "separable" parts of mathematics (namely, the theory of the natural numbers, including the principle of complete induction, together with whatever can be derived from it without the use of what Brouwer termed "axioms of existence") as independent of language and logic. The "pre-" part is intended to reflect the fact that they did not extend this attempt to find an extra-linguistic, extra-logical basis for knowledge of the continuum, a task which Brouwer took to be of critical importance. Cf. Brouwer (1951: 2–3).

4 Poincaré's Concern can also be seen as a problem concerning the compatibility of epistemic utility (i.e. the ability of proof to serve as a means of extending our knowledge) and what might be called the *logical conception of rigor*. According to this conception of rigor, concealment of assumptions in a proof is to be blocked by making every one of its steps or inferences so "small" as not to require any insight into the subject of the proof in order to verify it. Only in this way can one be assured that the steps do not conceal material assumptions concerning the subject that will then go undetected.

Making the steps of a proof too small, however, may compromise its epistemic utility. The clearest case, after all, of a fully "gapless" or rigorous proof is that of a circular argument, where the conclusion simply *is* one of the premises. Such proofs, however, are equally clear cases of epistemically useless reasoning. Thus, a minimal constraint on the epistemic utility of a proof is that it not be circular.

But though noncircularity is a necessary condition for epistemic utility, it seems not to be sufficient. It is doubtful, for instance, that one can turn a circular argument with a small number of simple premises into an epistemically useful one just by conjoining the premises and replacing them by the resulting conjunction. The argument obtained by such a procedure is not circular in the strict sense, since its conclusion is not literally the same proposition as any of its premises. But its epistemic utility is still doubtful, since one can clearly see that affirmation of the conclusion is part of what is needed for affirmation of the premise. Indeed, were one to "sharpen" the argument (i.e. to eliminate from the premises what is clearly unnecessary to its validity), one would end up with an argument whose only premise is the conclusion itself. As a general rule, it seems that any argument possessing recognized sharpenings that are circular is essentially lacking in epistemic utility.

5 Brouwer himself seems to have held a view something like this for the relationship of mathematics to natural science:

> The significance of mathematics with regard to scientific thinking mainly consists in this that a group of observed causal sequences can often be manipulated more easily by extending its of-quality-divested mathematical substratum to a *hypothesis*, i.e. a more comprehensive and more surveyable mathematical system. Causal sequences represented in abstraction in the hypothesis, but so far neither observed nor found observable, often find their realizations later on.
>
> (Brouwer 1948: 482)

Of currently greater significance to us at the present, however, is his idea that the logically driven classical conception of mathematics is related to genuine mathematics *in the same way* that, in the passage just quoted, mathematics is said to be related (or at any rate relatable) to natural science. Thus, logic-intensive classical reasoning is to genuine mathematical reasoning as mathematical of-quality-divested representational manipulation is to empirical investigation. They may be more or less accurate devices for predicting which causal sequences of intuitions will arise, but even when fully accurate they are not to be confused with the actual or potential intuitional verification of such sequences. More on this later.

6 Talk of "projection," of course, raises immediate questions concerning what it is that is the point or goal of the sorts of elementary constructional activities mentioned above. When engaged in those activities are we experimenting with the actual medium-sized physical objects of everyday experience in order to get a clear idea of the range of spatio-temporal manipulations through which they can be put? Or are we really attempting to effect, mentally, idealized operations (e.g. true reflections, true circumscriptions, true rotations, etc.) on ideal objects such as true planes, true circles, etc., and merely so designed that we are assisted in these tasks by our actual spatio-temporal fumblings with the geometrically imperfect physical objects of everyday life? Serious and interesting as these questions doubtlessly are, they are nonetheless not our concern here. For regardless of the true nature and subject of the constructional activity of elementary geometry, the difficulties involved in trying to extend it motivate one to develop a means of projecting it without extending it. It is this process of projection, and its possible relationship to the associated notion of extension, that are of primary interest to us here.

7 Heyting characterized the difference between his logic and classical logic as that separating a "logic of knowledge" from a "logic of existence." In a logic of knowledge, he went on to say, "a logical theorem expresses the fact that, if we know a proof for certain theorems, then we also know a proof for another theorem" (cf. Heyting 1958: 107). However, it is not clear that this way of thinking of epistemic incipience is at all close to the way in which Brouwerian epistemology conceives of it. For the Brouwerian, a given proof π' can only be said to be incipient in another proof π when the constructional activity or experience that is π is transformable into the constructional

activity that would be π'. The activity that is π and the activity that would be π', however, are different activities; and it would therefore not be correct to say that in doing the activity that is π one also *does* the activity that would be π'. From this and the fact that to know a proof for a theorem is to live or perform the activity that *is* that proof, it would seem to follow that in knowing the proof that is π one does not automatically know the proof that is π'. It thus seems that an intuitionist logic is not, as Heyting proposed, so much a logic of knowledge as a logic of knowledge-by-actually-doing.

It may also be that Brouwer regarded proof-activities as more robustly autonomous than did Heyting. Heyting saw proof-experiences as decomposable along contentual lines; that is, he held the view that if a proof-experience π had a compound proposition **p** as its content, then for each propositional component **c(p)** of **p**, there would be an isolable subactivity **c**(π) of π such that **c**(π) is a proof-activity whose content is **c(p)**. This, of course, suggests that proofs are logically deformable, and it is not clear that the kinds of proof-transformations that Brouwer had in mind (i.e. proof-transformations that constitute the optimal development – the free unfolding of – our mathematical knowledge) would have followed the lines of such deformation. He did admit that there were intuitionistic proof-activities corresponding to certain of the proofs constructed in an axiomatic system, but this may only have meant that they agreed in content and not in the compositional arrangement of subproofs. Indeed, though proof-activities for the Brouwerian can be structured, it is not likewise clear that the elements of that structure correspond to the subproof structure of an axiomatic proof, since there is no apparent reason why what structures some complex activity as an activity need follow the lines induced by contentual deformation. We shall return to these matters in the concluding section of the paper.

8 Both here and in the discussion to follow, we do not necessarily use the term "calculation" to signify the usual sort of effective, syntactical manipulation of symbols. Rather, we intend only the broader idea of a procedure that is something other than the literal thought or reasoning whose progress it (the so-called calculation) is supposed to chart. Thus, in calling logical reasoning "calculation," we are not intending to suggest that it is symbol-manipulation rather than contentual thought, but only that *as contentual thought* it is different from that contentual thought (namely genuine mathematical thought) of which it seeks to construct a "map."

9 We are not, therefore, denying that there is such a thing as "intuitionist logic," in the sense of a general theory of potential intuitionstic assertability. Rather, what we are denying is that the connections between propositions disclosed by such a theory are to be treated as constituting *rules of proof*; that is, rules that may actually be *used* in the construction of mathematical proofs. A connection between **p** and **q** established by intuitionist logic only tells us that from an experience of **p** we may expect to obtain an experience of **q** if we proceed in an appropriate way. But as Brouwer remarked (cf. 1948: 488), "expected experiences, and

experiences attributed to others are true only as anticipations and hypotheses; in their contents there is no truth," and propositions (i.e. the linguistic representation of possible contents of experience upon which the rules of logic operate) do not "convey truths *before* these truths have been experienced." What we take to be Brouwer's view of intuitionist logic is thus very different from that which is common today and which seems to have originated with Heyting.

It will probably be objected that the above fails to do justice to the fact that Brouwer often said such things as that having an algorithm for producing an equation is epistemically equivalent to actually producing it, and that I have therefore failed to realize that having an intuitionistic procedure for producing an experience and actually having that experience are to be treated as epistemologically equivalent. I plead innocent to this charge and have two replies to make in my defense. The first concerns correctly understanding what might be involved in equating the actual derivation of an equation with having an algorithm for deriving it. It is not clear to me that this is to be read as suggesting the equivalence of actually having an experience and being in a position practically to effect that experience. It might rather be read as saying that the experiencing of a computed equation simply *is* or *consists in* the possession of an algorithm for producing it. Taken in this way, the supposed equivalence of having an algorithm for producing and actually producing a result does not suggest any equivalence between actually having an experience and being in a position practically to effect that experience. It rather informs us of what sorts of things experiences of numerical results *are*.

The second point, which I shall only allude to here, is related to this matter of possessing algorithms. Let us suppose that having algorithms generally (and not just algorithms for producing numerical results) should be counted as having an intuitionistic experience (or something epistemically equivalent to it). Would it follow that the rules of intuitionist logic should be taken as devices for the construction of actual proofs? The answer, I think, is "no," for reasons that shall become clear in the concluding section of the paper.

10 This point was made repeatedly in Brouwer's writings.

> Will hypothetical human beings with an unlimited memory, who use words only as invariant signs for definite elements and for definite relations between elements of pure mathematical systems which they have constructed, have room in their verbal reasonings for the logical principles for tacking together mathematical affirmations? Or what comes to the same: Will human beings with an unlimited memory, while surveying the strings of their affirmations in a language which they use for an abbreviated registration of their constructions, come across the linguistic images of the logical principles in all their mathematical transformations? A conscientious rational reflection leads to the result that this may be expected for the principles of identity, of contradiction and of syllogism, but for the principium tertii

exclusi only in so far as it is restricted to affirmations about part of a definite, *finite* mathematical system, given once and for all whilst a more extensive use of the principle would not occur, because in general its application to purely mathematical affirmations would produce word complexes devoid of mathematical sense.... It follows that the language of daily intercourse between people with a limited memory, being necessary imperfect, limited and of insecure effect, even if it is organized with the utmost practically attainable refinement and precision, will only be suitable for its task of mnemotechnic, economy of thought and understanding in mathematical research and mathematical intercommunication, if any application of the principium tertii exclusi which is not restricted to a well defined system is avoided.

(Brouwer 1933: 443)

On account of the highly logical character of usual mathematical language the following question naturally represents itself: Suppose that an intuitionist mathematical construction has been carefully described by means of words, and then, the introspective character of the mathematical construction being ignored for a moment, its linguistic description is considered by itself and submitted to a linguistic application of a principle of classical logic. Is it then always possible to perform a languageless mathematical construction finding its expression in the logico linguistic figure in question? After careful examination one answers this question in the affirmative (if one allows for the inevitable inadequacy of language as a mode of description) as far as the principles of contradiction and syllogism are concerned; but in the negative (except in special cases) with regard to the principle of the excluded third ...

(Brouwer 1952: 141)

In the edifice of mathematical thought ... language plays no other part than that of an efficient, but never infallible or exact, technique for memorizing mathematical constructions, and for suggesting them to others; so that the wording of a mathematical theorem has no sense unless it indicates the construction either of an actual mathematical entity or of an incompatibility (e.g., the identity of the empty two-ity with an empty unity) out of some constructional condition imposed on a hypothetical mathematical system. So that mathematical language, in particular logic, can never by itself create new mathematical entities, nor deduce a mathematical state of things.

(Brouwer 1954: 523-4)

11 Of course it is true that "theorems" like the Continuity Principle, which may at first sight strike one as bounty on the side of the intuitionist, are *false* from the classicist's viewpoint, and hence scarcely to be regarded as an advantage. The same however, is true of

the "surplus" of the classicist. Judged from the intuitionist's vantage, it is not true, and hence not to be desired (as the intuitionist's critique of the soundness of classical logical reasoning makes clear).

12 Why not see the logical mastery of a scheme of implications as a type of practical knowledge? Surely the intuitionist would not want to deny that the extraction of logical implications is in some sense a form of mental activity. He must therefore hold the view that there is a basic difference between the mental activity that constitutes his mathematical construction-making and that mental activity which constitutes the extraction of logical implications. In the final analysis, of course, the Brouwerian will be obliged to offer some account of what that difference is.

13 It should perhaps be pointed out that characterizing logical analysis as mental activity does not automatically offer the classicist a way around Poincaré's Concern. For, as mentioned in the preceding note, that problematic could just as well be stated as a difference between two types of mental activity – logical and truly mathematical.

14 One might think of this as one way of sounding the Kantian theme which emphasizes the primacy of practical over theoretical reasoning.

15 Similar remarks can also be found in Heyting (1934: 14f).

16 Cf. Troelstra (1969: 6f.); Dummett (1977: 12f); van Dalen (1979: 133–4); McCarty (1983: 122–6).

17 Obviously, our statement of the assertion-conditions for the conditional and the univeral quantifier leave off the well-known "second clauses" (stating that the right-hand side has been proven) that Kreisel (1962) urged as necessary in order to insure the intuitionistic decidability of the intuitionistic notion of proof. This should not, however, be taken to suggest that we regard them as unnecessary, but only that their presence or absence is of no essential concern to the present discussion.

18 Some present-day authors who accept intuitionist logic are aware of some of the difficulties involved in treating intuitionistic proofs as mental processes rather than as syntactical entities, even though they do not seem to us to fully appreciate the consequences of what they are suggesting. Cf. Martin-Löf (1983, 1984, 1987) and Sundholm (1983).

19 Indeed, one might understand this as the source of the intuitionist's insistence on the construction of mathematical objects – without such construction, objects simply cannot be "given" to the mind in such a way as to allow genuine mental states (i.e. states having content). The interplay between "percepts" (= objects of construction) and "concepts" that such a view suggests may constitute another important respect in which intuitionist epistemology is Kantian in character. Since writing this paper, I have discovered a similar point in Tieszen (1992).

20 Cf. Brouwer (1948: 480–2). In addition to dropping this emphasis on the interiority or privacy of constructional mental activity, we also drop Brouwer's scheme of distinctions separating various causal-active and passive-reflective ways of extending one's mathematical experi-

ence. In his view, the epistemic quality (which may for him have really been a form of aesthetic quality) of the "unfolding" of the fundamental notion of two-ity is affected by the degree to which willful causal manipulation is involved in its production. "Shrewd" or "cunning" causal manipulation, whose aim is the "sinful" one of trying to control the stream of one's experience out of a preoccupation for one's own pleasure, produces experience of the lowest quality. Better quality results from a less calculating type of causal activity – termed "playful" by Brouwer – whose aim is to extend experience "without inducement of either desire or apprehension or vocation or inspiration or compulsion." There is "constructional beauty" in such playful causal activity, and it affords a "higher degree of freedom of unfolding" and greater "power, balance, and harmony" than shrewd causal manipulation. Still, playful causal unfolding of experience is inferior to the wholly free unfolding of experience, where one finds the "fullest constructional beauty."

REFERENCES

Bridges, D. and Richman, F. (1987) *Varieties of Constructive Mathematics*, London Mathematical Society Lecture Note Series, Cambridge: Cambridge University Press.

Brouwer, L.E.J. (1907) "On the foundations of mathematics" (doctoral thesis), in A. Heyting (ed.) *L. E. J. Brouwer: Collected Works*, vol. 1, Amsterdam: North-Holland, 1975, 11–101.

—— (1908) "The unreliability of the logical principles," in A. Heyting (ed.) *L. E. J. Brouwer: Collected Works*, vol. 1, Amsterdam: North-Holland, 1975, 107–11.

—— (1923) "On the significance of the Principle of Excluded Middle in mathematics, especially in function theory," in J. van Heijenoort (ed.) *From Frege to Gödel*, Cambridge, MA: Harvard University Press, 1967, 334–45.

—— (1933) "Volition, knowledge, language," in A. Heyting (ed.) *L. E. J. Brouwer: Collected Works*, vol. 1, Amsterdam: North-Holland, 1975, 443–6.

—— (1948) "Consciousness, philosophy, and mathematics," in A. Heyting (ed.), *L. E. J. Brouwer: Collected Works*, vol. 1, Amsterdam: North–Holland, 1975, 480–94.

—— (1949a) "The non-equivalence of the constructive and the negative order relation on the continuum," in A. Heyting (ed.) *L. E. J. Brouwer: Collected Works*, vol. 1, Amsterdam: North-Holland, 1975, 495–7.

—— (1949b) "Contradictority of elementary geometry," in A. Heyting (ed.) *L. E. J. Brouwer: Collected Works*, vol. 1, Amsterdam: North-Holland, 1975, 497–8.

—— (1951) *Brouwer's Cambridge Lectures on Intuitionism*, ed. D. van Dalen, Cambridge: Cambridge University Press, 1981.

—— (1952) "Historical background, principles, and methods of intuitionism," in A. Heyting (ed.) *L. E. J. Brouwer: Collected Works*, vol. 1, Amsterdam: North-Holland, 1975, 508–15.

—— (1954) "Points and spaces," in A. Heyting (ed.) *L. E. J. Brouwer: Collected Works*, vol. 1, Amsterdam: North-Holland, 1975, 522–48.

—— (1955) "The effect of intuitionism on classical algebra of logic," in A. Heyting (ed.) *L. E. J. Brouwer: Collected Works*, vol. 1, Amsterdam: North-Holland, 1975, 551–4.

Dalen, D. van (1979) "Interpreting intuitionist logic," *Proceedings, Bicentennial Congress, Wiskundig Genootschap*, vol. 1, Mathematical Center Tracts, Mathematisch Centrum, Amsterdam.

Dummett, M. (1977) *Elements of Intuitionism*, Oxford: Oxford University Press.

—— (1978) "The philosophical basis of intuitionist logic," in *Truth and other Enigmas*, Cambridge, MA: Harvard University Press.

Heyting, A. (1934) *Mathematische Grundlagenforschung. Intuitionismus. Beweistheorie.* Page references are to the 1974 reprint of this by Springer-Verlag.

—— (1956) *Intuitionism: An Introduction*, Amsterdam: North-Holland. Page references are to the 3rd revised edition of this work published in 1971.

—— (1958) "Intuitionism in mathematics," in R. Klibansky (ed.) *Philosophy in the Mid-Century*, Firenze.

—— (1975) *L. E. J. Brouwer: Collected Works*, vol. 1, Amsterdam: North-Holland.

Kreisel, G. (1962) "Foundations of intuitionist logic," in E. Nagel, P. Suppes, and A. Tarski (eds) *Logic, Methodology and Philosophy of Science*, Amsterdam: North-Holland.

Lorenzen, P. (1984) *Elementargeometrie. Das Fundament der analytischen Geometrie*, Mannheim.

—— (1985) "Das Technische Fundament der Geometrie," *Philosophia Naturalis* 22: 22–30.

McCarty, C.D. (1983) "Introduction," *Journal of Philosophical Logic* (Special Issue on Intuitionism) 12: 105–49.

Martin-Löf, P. (1983) "On the meanings of the logical constants and the justification of the logical laws," *Atti Degli Incontri di Logica Matematica* 2: 203–81.

—— (1984) *Intuitionistic Type Theory*, Naples: Bibliopolis.

—— (1987) "Truth of a proposition, evidence of a judgment, validity of a proof," *Synthese* 73: 407–20.

Poincaré, H. (1902) "Science and hypothesis" in G. Halsted (ed. and trans.) (1946) *The Foundations of Science*, Lancaster, PA: The Science Press.

—— (1905) *The Value of Science*, in G. Halsted (ed. and trans.) *The Foundations of Science*, Science Press, 1946.

Sundholm, G. (1983) "Constructions, proofs, and the meanings of the logical constants," *Journal of Philosophical Logic* 12: 151–72.

Tieszen, R. (1991): "What is a proof?," in M. Detlefsen (ed.) *Proof, Logic, and Formalization*, London: Routledge, ch. 3.

Troelstra, A. S. (1969) *Principles of Intuitionism*, Lecture Notes in Mathematics 95, Berlin: Springer-Verlag.

INDEX

a posteriori knowledge 35, 42; truth 41
a priori 13; belief (as propositional precondition of thought) 73, 74, 75, 76, 79–82, 105; judgement 36; knowledge 2, 12, 35, 42; propositions (as mind-imposed) 74, 80, 82; truth 33, 34, 38, 40, 41, 43–5, 61, 63; warrant 73–6, 80–2, 84, 105
abstract objects 33
abstraction 104
algorithms 198–200, 202, 246
Allison, H. 51, 52
Amit, D. 170
analysis: Cauchy–Weierstrass foundation of 203; classical 142; nonstandard 16, 142, 149
analytic *a priori* 70
analytic truth 152
analytic truths of geometry 63
analyticity 99, 100; Kantian test for 70
antanairesis 137
anthyphairesis 137
anti-foundationalism 171, 177, 184
antirealism 33, 34
aprioricity 217, 218
apriorism 83, 84, 90, 91; cognitive 84, 85, 89; communal 87, 89
apriority 35–8, 46, 62
Archimedean axiom 136
Archimedean ordering 138, 139, 150

Archimedes 101
Aristotle 39, 156, 185
Augustine 34
axiomatic system, truth in 156

Becker, O. 156
Benacerraf, P. 30, 101, 104
Berkeley, G. 160
Blackburn, S. 205
Bolzano, B. 16, 66; proof of the intermediate value theorem 181
Bolzano–Weierstrass theorem 181
Boolos, G. 182, 203
Borel, E. 243
Brittan, G. 63
Brouwer, L.E.J. 63, 145, 199, 208, 210, 214, 216, 217, 240–9; epistemology of mathematics 218, 222–33; first critique of classical logic 223–5; on inaccuracy of classical logic 226, 227; second critique of classical logic 225–7
Byrd, M. 242

calculation 118–20, 124, 125, 225, 226, 245; arithmetical 102, 103
Cantor, G. 16, 90, 95
Cantor's Paradise 178
Carnap, R. 66, 78
Cauchy, A. 95, 142
chess 211
Chomsky, N. 69
Church's Thesis 198, 201, 204, 205

Cocchiarella, N. 203
Coffa, A. 181
Columbus, C. 107
common notions (Kant) 54, 58, 63
completeness 4, 184, 188–91;
    axiom of 136
computability theory 198–201
conformity, the issue of 184, 187, 189, 192
congruence 136, 139
constructions, geometrical 58, 59, 135, 143
constructivism 71, 72, 75, 78
Continuity Principle, The 227, 247
coordinate system 140
Corcoran, J. 205
counting 101–4, 114
Cummins, R. 106

Dalen, D. van 248
de dicto/de re ambiguity 45, 47
Dedekind, R. 16, 50, 65, 66, 69–72, 74–9, 81, 82, 90, 94–7, 101, 104, 105, 140
Dedekind-cuts 136
Dedekind–Peano axioms 11, 99
deduction, Cartesian 68
deductivism 8
demonstration 136; geometrical 135, 143
demonstrative knowledge 39
demonstrative science 39
Desargues' Theorem 138, 146, 151, 152
Descartes, R. 98, 101, 140, 174
Detlefsen, M. 31, 199, 203; and Luker, M. 123–5, 134
diagrams (and proof) 60
Diamond, C. 169
dianoia 41
Dingler, H. 135, 148
Dirac δ-function 160
Došen, K. 242
Drei-Platten-Verfahren 148
Dummett, M. 238, 248

Einstein, A. 155, 156, 159
Einstein synchronization 155
Elizur, S. 170

Erkenntnisse 36
Euclid 58, 59, 63, 64, 136, 138, 140, 156, 175, 183, 204;
    algorithm of 137; theorem of 6, 9, 13–15, 28–30
Eudoxus 140; theory of proportions 138, 139, 141, 142
Euler, L. 159, 162, 168
explanation 68

Fermat principle 168
Feynman integral, the 160, 161, 168; and mathematical rigor 161
Feynman, R. 166
fictionalism 88
Field, H. 30, 31
first-generation logicism *see* logicism, first-generation
Foley, R. 67
formal analogy (of the physicist's symbolic manipulations to mathematics) 161–4, 166, 167; and Euler's manipulation of power series 162
formal instructions 143–8, 153; norming of 143, 144
formalism 8, 10, 147, 156
formalization 180
Formprinzip 156
forms, ideal geometrical 143
foundationalism 4, 67
Four-Color Theorem, the 123
Fowler, D.H. 156
Fraassen, B.C. van 168
Frege, G. 11, 12, 16, 38, 41, 44, 50, 63, 65, 66, 69–72, 74, 75–9, 82, 90, 94–7, 100, 104, 105, 107, 175, 182, 184, 204
Friedman, M. 47, 49–51, 54, 56, 58, 60, 63, 78
Fritz, K. v. 156
Furstenburg, H. 170

Γ-function, the 163, 168
Gauss, K.F. 124, 125, 154
Gell-Mann, M. 158, 167
geometry: analytic 136, 137, 143, 145, 151; applied 47, 61; arithmetization of 137;

INDEX

axiomatic 145; axiomatized 136, 137; constructive 147; elementary synthetic 135, 137, 143, 146, 150; empirico-constructional conception of 221, 222; Euclidean 42, 43, 56, 58, 61, 76, 151, 152, 154, 155, 159; non-Euclidean 51, 153; projective 51; pure 47; Riemannian 154; synthetic 151
Gersten, J. 170
God 62
Gödel, K. 68
Gödel's Second Theorem 176
Gödel's theorems 65, 97, 198
Gödel-Cohen consistency theorems 88
Goldfarb, W. 175, 176, 203, 204
Goodman, N. 203

Heath, T, 31
Heijenoort, J. van 175, 203
Hempel, C. 166
Heyting, A. 234, 237, 244-6
hidden contradictions: Turing on 166; Wittgenstein on 166
Hilbert, D. 11, 43, 50, 51, 58, 135, 145, 156, 169, 178, 184
Hilbert's Program 171, 174-7, 204
Hintikka, J. 56, 203, 204
historicity 89, 90, 95
Hobbes, T. 198, 199
Hume, D. 165
hypothetico-deductive justification 89, 93

ideal justification 185, 189, 193-6, 198, 201, 202
immanent realism (about mathematical objects) 8, 9, 10, 29, 30; and the immanent conception of truth 8
inconsistency, mathematical 3, 164-6; and accuracy in physics 166
inference: experience-intensive view of 199, 201, 209, 223, 224; explanatory 3, 68, 69; ideally rational 4; logic-intensive (representation-intensive) view of 199, 209, 219-22, 224; logical 1-4, 49, 208, 209; mathematical 1, 57
infinitesimals 16, 141, 160
Inhetveen, R. 156
instrumentalism 88
intermediate value theorem 181
intuition: Cartesian 68; formal 48, 55; immediate 59; nonempirical 55; pure 37, 48, 51, 54, 56, 58, 60, 61; sensible 48, 55, 57, 61; space as a form of 47, 48; space as formal 47, 48
intuitionism 8, 71, 172, 203, 208-49 *passim*; First Act of 223
Isaacson, D. 242

Janich, P. 155, 156

Kambartel, K. 242
Kant, I. 2, 3, 33, 34, 36-40, 44-50, 52-8, 60-2, 70, 71, 135, 145, 152, 153, 204, 242, 248; theory of geometry 59
Kazhdan, D. 170
Kepler, J. 159
kinetics, general 135
Kitcher, P. 31, 35, 65, 68, 69, 72, 74, 76-82, 84, 87-96, 101, 105, 106, 168, 181, 203
knowledge, social conception of 67
Kolchin, E. 158
Koski, E. 107
Kraut, R. 203
Kreisel, G. 110, 133, 204, 205, 242, 248
Kripke, S. 35

Lakatos, I. 106, 159, 167
language, symbolic 156
Lebesgue, H. 243
Leibniz, G.W. 16, 33, 38-46, 48, 52-4, 60, 62, 63, 66, 141, 142, 153, 156; *calculus ratiocinator* 175; *lingua characteristica* 175, 198
Levy, I. 166, 170

253

light, propagation of 153, 155
logic: Aristotelian 191, 205; as calculus 175, 176, 198; and computation 197–203; first-order 3, 4, 66, 67, 86, 182, 191, 204; higher-order 193, 203; intuitionist 233–40, 245; as language 175, 176; ω- 204; propositional 191, 205; quantum 106; rationalist conception of 3, 67, 183–5, 188, 192–5, 201, 205; second-order 4, 191, 204; semantic conception of 3, 183, 185–8, 196, 197, 202, 205
logical consequence 67
logical reasoning *see* reasoning
logical truth 67
logicism 65–104 *passim*; 171, 174, 176, 177, 179–81, 204, 205; first-generation 65, 66, 70–2, 97
Lorentz transformation 155, 156
Lorenzen, P. 156, 221
Luce, L. 203

McCarthy, T. 105
McCarty, C.D. 203, 248
McCord, G.S. 31
Mac Lane, S. 31
Maddy, P. 26, 27, 31, 68, 92, 93, 103, 106, 203
Malkin, N. 170
Manin, Y. 161, 168
Margalit, A. 170
Martin-Löf, P. 238, 248
mathematical induction 21
mathematical knowledge: classical view of 222; extension of 214–16; local character of 232; vs logical knowledge 210; modal construal of 213–15; as practical knowledge 229–31; subjective construal of 213, 214; theoretical conception of 229, 230
mathematical naturalism 82, 89, 90
mathematical objects 7–9, 11, 16, 17, 29; contact with 17, 25, 27
mathematical realism *see* realism

mathematical reasoning *see* reasoning
mathematical rigor 4, 95, 96, 158, 159, 167, 212; and proof 158; and twentieth-century physics 160, 161, 164–6
mathematical structures 18
mathematics: Brouwerian epistemology of 222–33; classical epistemology of 217–22, 225, 228; idealization in 151
mechanism 201
mentalism 145
Mill, J.S. 22, 114
mirror-mappings 135
moderate-foundationalism 173, 174
Moore, G. 203
Morgenbesser, S. 170

naive comprehension 12, 103
naturalism 167
necessity 35, 37, 38, 45, 46, 62
Ne'eman, Y. 169, 170
Neumann, J. von 101, 186
Neurath, O. 197
Newton, I. 16, 42, 105, 141, 159, 160
noesis 39, 41, 43
nominalism 24
nonexperiential cognition 87, 89, 94, 98
numbers: irrational 137; nonstandard 142; real 136, 141

observation 145
ω-logic *see* logic, ω-
ω-rule 204

P-foundationalism 174
parallel postulate 136, 150
Parsons, C. 42, 105, 204
Pasch, M. 51
patterns 18, 22, 23, 25–9; as posits 25
Peano, G. 136
Peirce, C.S. 65
Pitowsky, I. 170
Plato 33, 39, 40, 42, 43, 46, 98,

143, 145; conception of geometry 40, 41; doctrine of forms 41
platonism 24–6, 34, 145
Poincaré, H. 179, 181, 182, 204, 208, 209, 231, 233, 242; on mathematical induction 213
Poincaré's Concern 210–17, 241, 243, 248
Popper, K. 221
pragmatism 79, 80
Principle of Noncontradiction 52, 53, 60
Principle of Sufficient Reason 38
Principles of Pure Understanding 44, 48
proof: classical view of 217, 218; the creative aspect of 111, 113, 128, 129; vs experiment 111, 114–16, 118, 126; as a giver of standards 111, 114, 115, 126–8, 132; and perspicuity 110, 116, 122, 127, 132, 133; self-sufficiency of 124; and surveyability 111, 115, 116, 123, 127, 133; as type and as token 110, 117, 125–8, 130–2; as a variable, practical phenomenon 117, 118, 132
proof, discursive (vs demonstration) 49
proof-construction 114, 123, 126, 127; vs correct experiment 116–21; and imagination 118, 119; intuitionist conception of 234, 235, 238–40
proofs: and their connection with mathematical objects 7; contextual character of 15, 16; deductive 142; dot 19–21, 26; uses of 11; working 12–16
psychologism 178–81
Putnam, H. 30, 68
Pythagorean program 137
Pythagorean theorem 135, 137, 151
Pythagoreanism 141, 169

quantum electrodynamics (QED) 164–6, 169

Quine, W.V.O. 13, 30, 31, 34, 42, 43, 62, 65, 66, 68, 91, 96, 97, 99, 106, 166
Quinn, P. 242

rationalism 171, 200, 201
realism: mathematical 7–9, 33, 34; transcendent 8, 9, 26; *see also* immanent realism
reasoning: logical 210–12; mathematical 210–12; nonrigorous 159, 161, 163–5
reasoning and computation 199–203; strong equation of 200; weak equation of 200, 201
relativity theory 155
Resnik, M. 180, 203
Riemann ζ-function, the 162
rigor, mathematical *see* mathematical rigor
Robinson, A. 142, 160
rules: formal 3, 148; semantic 3, 148
Russell, B. 35, 42, 49–52, 54, 62, 66, 69, 133, 179, 181; and Whitehead, A.N. 13, 68, 182
Russell's paradox 65, 74, 97, 103, 182

Schirn, M. 105
Scholz, B. 203
Schumm, G. 203
self-evidence 183–5
Shanker, S. 110, 116, 132, 133
Shapiro, S. 31, 105, 242
Silverman, A. 203
Simon, B. 170
Skolem, T. 177; "opportunistic" option 177
Sluga, H. 105
soundness 184, 188–91
Stein, H. 105, 106
Steiner, M. 31, 106
Stekeler, P. 242
Sternberg, S. 170
strong-foundationalism 173
sufficiency, the issue of 184, 187, 189, 192
Sundholm, G. 248

surveyability 111, 115, 122
synthetic *a priori* 44, 151, 152, 212;
   propositions of geometry 57
synthetic truth 152
synthetic truths of geometry 63

Tait, W. 30, 175
tangrams 139
Tarski, A. 8, 9, 31, 185
templates 1, 18, 19, 22–4, 26, 27;
   conforming to 18
Thales, theorem of 135, 151
theory/metatheory distinction 175
Tieszen, R. 248
transcendent realism *see* realism
Troelstra, A. 248
truth: immanent conception of 8;
   semantic concept of 142
truth/proof problem 7, 9, 10, 17, 23, 24, 27, 29; platonist solution to 24; reformulation of 10
Turing, A.M. 166
Turnbull, R. 203
Tymoczko, T. 123, 134

universality, strict 37

Veblen, T. 11
violations of rigor and success in physics: illustrated by a calculation with analytic functions 162–4; illustrated by calculation of perturbation due to vacuum polarization 164–6

Waerden, B. van der 31
Wagner, S. 183, 185, 191, 202, 205
Weierstrass, K.T. 95
Wiener integral, the 168
Weyl, H. 172
Wilder, R. 31
Wittgenstein, L. 2, 10, 34, 44, 65, 110, 133, 166, 169, 199, 205; his apriorism concerning proof 110, 111, 122, 123, 129, 130, 131; on the limits of empiricism 113, 114; on mathematical discovery 127; philosophy of mind 119; on proofs and experiments 110–12, 114, 115, 118–21, 126; relative importance of logical vs epistemic properties to his philosophy of mathematics 112, 113, 116; unconditional nature of mathematical standards 113, 115
Wolff, C. 48
Wright, C. 205
Wright, G.H. von 242

Zalcman, L. 168, 170
Zemach, E. 170
Zermelo, E. 43, 76, 101, 106, 136, 199
Zermelo–Fraenkel set theory 43, 105

Printed in Poland
by Amazon Fulfillment
Poland Sp. z o.o., Wrocław